MAMMOTH CAVE

Mammoth Cave Curiosities

A Guide to Rockphobia, Dating, Saber-toothed Cats, and Other Subterranean Marvels

COLLEEN O'CONNOR OLSON

Scholarly publisher for the Commonwealth,
serving Bellarmine University, Berea College, Centre College of Kentucky, Eastern Kentucky University, The Filson Historical Society, Georgetown College, Kentucky Historical Society, Kentucky State University, Morehead State University, Murray State University, Northern Kentucky University, Transylvania University, University of Kentucky, University of Louisville, and Western Kentucky University.

Editorial and Sales Offices: The University Press of Kentucky
663 South Limestone Street, Lexington, Kentucky 40508-4008
www.kentuckypress.com

Unless otherwise noted, photos are by Rick Olson.

Cover images: *Inset:* magnifying glass © NiKolayN, courtesy iStock; saber-toothed tiger © brackish_nz, courtesy iStock. *Background, clockwise from top left:* photo by Rick Olson; © nicoolay, courtesy iStock; © benkrut, courtesy iStock; © salajean, courtesy iStock; © nicoolay, courtesy iStock.

Library of Congress Cataloging-in-Publication Data

Names: Olson, Colleen O'Connor, 1966- author.
Title: Mammoth Cave curiosities : a guide to rockphobia, dating, saber-toothed cats, and other subterranean marvels / Colleen Olson.
Description: Lexington, Kentucky : University Press of Kentucky, 2017. | Includes bibliographical references and index.
Identifiers: LCCN 2016046791 | ISBN 9780813169255 (pbk. : alk. paper) | ISBN 9780813169279 (pdf) | ISBN 9780813169262 (epub)
Subjects: LCSH: Mammoth Cave (Ky.)—History—Anecdotes. | Mammoth Cave (Ky.)—Anecdotes. | Curiosities and wonders—Kentucky—Mammoth Cave—Anecdotes.
Classification: LCC F457.M2 O445 2017 | DDC 976.9/754—dc23
LC record available at https://lccn.loc.gov/2016046791

This book is printed on acid-free paper meeting the requirements of the American National Standard for Permanence in Paper for Printed Library Materials.

Manufactured in the United States of America.

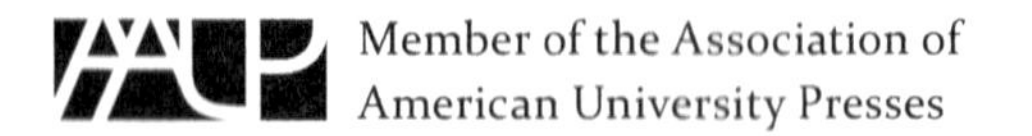

To the Mammoth Cave Guide Force,
a dedicated bunch of people
who share their knowledge and love
of the world's longest cave
with park visitors every day

Contents

Introduction

Elton John, eyeless fish, Twinkies, Ralph Waldo Emerson, and carbon-14 dating usually don't come up in the same conversation—unless you're at Mammoth Cave, where unusual topics are normal.

Working as a cave guide for twenty-plus years, I have encountered plenty of fascinating topics that were

- Difficult for me and other ordinary people to understand, such as cosmogenic isotope dating.
- Not written down, such as President Reagan's Secret Service agents mistaking a bicycle tire blow-out for a gunshot.
- Scattered but in need of being brought together, such as works by Herman Melville, Jules Verne, and L. Frank Baum that mention Mammoth Cave.
- Of great interest but not written much about, such as the effect of earthquakes on caves.
- Seemingly trivial but really important, such as what original Twinkies, beer, drywall, and Mammoth Cave have in common.

To find some answers, I read scientific papers and asked park scientists to explain them to me; interviewed current and former cave guides; communicated with archaeologists, cavers, earthquake experts, geologists, paleontologists, brewers, bakers, war historians, and army cooks; and combed through historic letters, newspaper articles, journals, court records, and books.

This quest resulted in several papers I wrote for National Park Service staff. After keeping this material to ourselves for several years, I updated it for the public, and it is now available to all park visitors, cavers, and other fans of Mammoth Cave in this book.

1

The Mammoth Cave Dating Guide

(Not *How to Get a Date with a Mammoth Cave Guide*)

With Rickard A. Olson and
Rickard S. Toomey III

How old is Mammoth Cave? How about the limestone, the passages, stalactites, and prehistoric artifacts? How do scientists figure out the age of the cave and things in it? Here are the basics on common dating techniques used at Mammoth Cave.

Handy Definitions

Relative Dating. Relative dating refers to techniques that show whether an object is older, younger, or the same age as something else but not the exact age of the object. For example, rock layers at the bottom of Mammoth Cave are older than layers at higher levels because the higher rock layers were deposited on top of the lower layers. (Exceptions can happen when tectonics or faulting cause rock layers to slide or flip, putting older layers on top of younger layers.)

Absolute Dating. Absolute dating shows how many years old something is. To get an absolute date, scientists must have something reliable to measure, such as a radioactive isotope whose half-life they know.

Isotope. When two or more atoms of the same element (such as carbon, oxygen, or lead) have the same number of protons and electrons but a different number of neutrons, they are still the same element but different isotopes because of the difference in the number of neutrons. Some isotopes are stable and don't change; others are radioactive, which causes them to decay. For example, carbon-12 (C12) and carbon-13 (C13) are stable, but carbon-14 (C14) is radioactive; all are carbon, but they are different isotopes.[1]

Half-life. Radioactive isotopes decay at a certain rate. The amount of time it takes half of the atoms in a particular isotope to decay is the isotope's half-life.

How Old Is It, and How Do We Know?

Prehistoric Artifacts

The oldest artifacts in Mammoth Cave are about 5,000 years old, but most are between 2,200 and 2,800 years old.[2] *Radiocarbon dating,* also called *C14* or *carbon dating,* is used to date artifacts made of plant or animal remains. Living things contain three carbon isotopes, C12, C13, and C14. When the plant or animal dies, the C14 begins to decay into nitrogen-14. C14 has a half-life of 5,730 years,[3] so after 5,730 years half of the C14 will have decayed. After another 5,730 years, half of what's left will decay, and so on. There eventually won't be enough C14 left to measure, so the technique is good only for artifacts up to about 50,000 years old.[4]

Formations

How quickly a stalactite or other formation grows depends on how quickly the mineral calcium carbonate is being deposited on it. Conditions differ from place to place and even change

Carbon dating shows this ancient climbing pole to be about 2,000 years old.

with time at the same place. You can't tell how old a formation is by its size, but geologists can learn an approximate date using *uranium-series dating.* Scientists used uranium-series dating to determine that a stalagmite from a cave in Yemen had a

You can't tell how old this cave formation is by its size, and it's too beautiful to break off a piece to test in the lab. (Courtesy National Park Service)

growth rate varying from one inch in 40 years to one inch in 317 years—an average of 90 years per inch in this particular case.[5] Growth rate differs a great deal from one formation to another. An 8-inch-long stalagmite from Diamond Caverns (a show cave near Mammoth Cave) was dated to be about 306,000 years old at the base and about 169,000 years old at the tip. The Diamond Caverns stalagmite grew a little more than 1¼ inches in 20,000 years, which is a growth rate of one inch per 16,000 years![6] Even though formations are not usually cut in the park, some formations from Mammoth Cave have been uranium-series dated to be more than 350,000 years old.[7]

Here's how uranium-series dating works. Groundwater contains some uranium-234, which gets deposited in formations.[8] Uranium-234 has a half-life of 245,000 years; as it decays, it becomes thorium-230. The thorium-230 also decays; it has a half-life of 75,000 years. Scientists can look at the ratio of uranium-234 to thorium-230 to learn when the water deposited the calcium carbonate and the uranium. This technique works for formations up to 600,000 years old.[9]

As the decay process continues, the thorium becomes radium, and the radium decays into radon—a gas that can cause cancer with high, long-term exposure. A few hours in the cave won't affect you, but you don't want radon in your house.

Passages

Cave passages are just space. You can't date empty space, but scientists can determine approximately how long a passage has been dry by using *cosmogenic isotope dating* on quartz sand or gravel in the sediment. This technique provides only a minimum age estimate on when a passage formed because passages began developing long before the last sediments were washed in.

Here's how cosmogenic isotope dating works. Scientists look at the ratio of aluminum to beryllium in quartz pebbles. Quartz is made of silicon and oxygen. Cosmic rays from outer space bombard quartz that's above ground, causing the silicon to be converted into aluminum-26 and the oxygen into beryllium-10. Aluminum-26 is produced six times as fast as beryllium-10, so most rocks above ground have a six-to-one ratio. If quartz gets washed into a cave, it is no longer bombarded with cosmic rays, so the aluminum and beryllium stop being produced; they just decay. Aluminum-26 has a half-life of 700,000 years, and beryllium-10 has a half-life of 1.5 million years. Because they decay at different rates, the ratio of aluminum to beryllium changes with time. Because we know the ratio and rate of decay of aluminum-26 and beryllium-10, we can tell how long the pebble has been in the cave out of the sun. When the passage is no longer in the flood zone (or other wet place, such as a shaft), pebbles no longer get washed in. So much of the pebble's aluminum and beryllium will eventually have decayed that there's not enough left to measure, so this technique is

How do you date this passage? It's just a lot of space!

good only for rocks that have been buried in the cave less than 5 million years.[10]

Samples from the cave rooms called Methodist Church and Violet City show that it has been about 2.27 million years since large sediments were washed into those areas. Samples from Forks of the Cave show that about 1.73 million years ago a collapse in Kentucky Avenue caused water to be diverted into Rose's Pass and Boone Avenue.[11] That's how long it has been since gravel was washed in; the passages started forming much earlier.

Bedrock

Unlike stalagmites or quartz gravel in a cave passage, the bedrock of Mammoth Cave cannot be dated directly because limestone typically lacks enough uranium or other radioactive elements to date. So the way geologists determine the age of sedimentary rocks such as limestone is to correlate fossils in the rock to fossils in other places where there is also a layer nearby that can be dated by a radioactive decay clock method.

For example, fossil algae and foraminifers (single-cell protozoans that build shells) are found both in Mammoth Cave's St. Louis Formation limestone[12] and in limestone strata near Guangxi, South China.[13] (In cave talk, the word *formation* often refers to a decoration such as a stalactite, but in names such as "St. Louis Formation" it refers to a specific section of bedrock.) Also, there are fossil species of conodonts (small marine carnivores identified mostly by their teeth) in a limestone layer in southern Illinois that is the same as the Paoli limestone in Mammoth Cave.[14] Conodonts of the same age are in rock in Harz, Germany; fortunately, clay layers nearby have been dated using a uranium-lead radioactive decay clock method.[15] The age of the German limestone was determined to be 334 million years old. Thus, because of the common fossil assemblage, we know that Mammoth Caves Paoli limestone is approximately the same age.

Special thanks to Rickard A. Olson, ecologist at Mammoth Cave National Park; Rickard S. Toomey III, director at the Mammoth

Cave International Center for Science and Learning; Dr. Jeff Dorale of the University of Iowa; and Dr. Darryl E. Granger of Purdue University. I could not have written this chapter without their help.

2

Rockphobia

Rockfall, Earthquakes, and Caves

Mammoth Cave visitors often ask if earthquakes occur in Kentucky. It's not a fascination with tectonic movement that drives this question; it's a fear of falling rocks. Should people worry about rockfall and earthquakes in the cave?

Rockfall without Earthquakes

True personal story: One busy summer day, a visitor asked me if rocks ever fall in the cave. While I explained the rarity of rock fall, I saw a small rock drop in my peripheral vision. The group was so mesmerized by my talk (at least I like to think I mesmerized them) that no one said anything about the falling rock. I wondered if someone had tossed it as a joke. When the group left, I picked up the rock. It looked like a fresh break, not dirty like a rock that someone picked up off the ground.

True personal story number two: While I was walking out of Mammoth Cave's Violet City Entrance, a small rock fell off the ceiling and hit me on the left shoulder.

I have kept a running tally of all the rocks I know of that have fallen in Mammoth Cave since I started working at the cave in 1992. As of June 2015, I counted about fifty rocks that had fallen on cave tour routes. This is by no means all the rocks that fell in that time, just the number I saw on trails or that my fellow guides told me about—not exactly a scientific study.

Rocks don't often fall in the cave. Being hit by a rock is rare enough to make it not worth worrying about; you're more likely to be hit by a falling branch in the woods. Rockfall is less common in the cave than above ground because underground you get away from tornados, ice, and most other weather, but you can't get away from gravity. Rocks fall in the cave, they fall on tour trails, and if you're in the wrong place at the wrong time, they can fall on you.

The Big Rock Fall in 1994

A big ice storm caused the park to close for a few days in January 1994. When the park reopened, the guide brought the first tour into the cave and walked through the room called the Rotunda toward where guides often stood on the trail by the handrail and a light switch, but he saw no trail, no rail, and no light switch. An estimated forty tons of rock[1] had fallen on the trail and part of the historic saltpeter works—the biggest rockfall in the recorded history of Mammoth Cave.

Park guides and scientists were surprised, but not too surprised. We had never seen a rockfall that big before, but the domed ceiling that gives the Rotunda its name indicates that plenty of rocks have collapsed there over time. Being near the Historic Entrance, the Rotunda gets seasonal temperature changes that cause some expansion and contraction, which increase the chance of rockfall. The temperature outside the cave on January 19, 1994, was –8.6°F; the cold air blew into Rotunda, causing thermal contractions.[2] Cold winter air has always blown through Historic Entrance because it's a natural entrance, and human modifications to it over the years have increased the amount of air that blows in.[3] Once you get away from the entrance and the weather, the likelihood of rockfall decreases.

Earthquakes and Caves

I contacted world-renowned hydrologist, geologist, and caver Dr. Art Palmer to ask what he knows about earthquakes and

Much of the cave escapes the weather, but cold winter weather increases the possibility of rockfall near entrances to the cave. (Courtesy National Park Service)

caves. He wrote, "I'm not sure we can find a definitive statement, but the general gist is [that] surface waves are the most destructive (they involve much more motion than waves that go through the ground), and they die out with depth, just as waves on the ocean do. Most caves are deep enough that the amount of motion caused by surface waves will be negligible, and the energy from waves that do penetrate to that depth is fairly small." He explained that rock moves as a coherent mass, so it is mainly thick sediment at the surface that is prone to disruptive shake, like pudding. Walls and ceilings of caves are pretty coherent and shake in unison. Piles of breakdown would be susceptible to disruption, but there are no eyewitness accounts of anything more than a couple of loose rocks tumbling down preexisting breakdown piles during earthquakes and rarely even that.[4]

I also wanted to hear from an earthquake expert, so I contacted U.S. Geological Survey seismologist Dr. Lucy Jones (a.k.a. the Earthquake Lady). She wrote, "Seismologists have derived the equations of motion for earthquake waves, and they clearly show that ground motions on the surface of the earth are larger than underground. We have many places where we measure the ground motion at the surface and in a borehole at some depth below the surface, and they all show that the surface motions are larger. So caves and tunnels are safer places to be—but it is pretty hard to convince most humans of that fact."[5]

In a Cave during an Earthquake

Some firsthand accounts from individuals who have been in a cave during an earthquake make me feel safe.

Mammoth Cave natural-resources specialist Brice Leech told me about two separate earthquakes he experienced at Wind Cave in South Dakota.

> I was crawling through a two-foot-tall passage not far off the tourist trail, alone, when the rocks began to rumble for several seconds. I immediately stopped and braced myself (I knew it was an earthquake) for whatever was going to happen. Then suddenly I remembered my train-

ing and knew I would not feel or see anything strange, at which time I just laid back and enjoyed the music the quake gave me. The total time that elapsed was probably fifteen to twenty seconds. Upon my return to the surface, I was asked if I had felt the quake. Of course, my answer was no.

Another time on a water-collection trip I had an assistant. This particular time occurred while walking the tour trail. Upon the beginning of the quake, my partner responded with panic, asking what the noise was. My response was that it was a semi driving the road on the surface, which calmed his nerves. Upon exiting the cave, I corrected my earlier statement, at which time he responded with excitement. The time lapse of the quake was approximately the same as before, fifteen to twenty seconds.

Feeling the quake, seeing dust rise and fall . . . rocks falling from the ceiling and walls [were] not experienced.[6]

Mike Wiles, the cave management specialist at Jewel Cave, shared his story.

I have been in Wind Cave and Jewel Cave during an earthquake at least three times. Both times at Jewel, we felt nothing but heard what sounded like a distant rumbling. . . . We all asked, "Did you hear that?" However, none of us felt any vibration. I checked with South Dakota School of Mines and Technology and found that there had been a 4.0 earthquake at that exact time, somewhere between Jewel Cave and Edgemont, South Dakota.

At Wind Cave, it sounded like distant wind whistling through a constriction. Some friends who lived near Cascade Springs felt the quake on the surface. It rattled the windows and made doors swing open or closed, so they were worried about us—but we felt nothing.[7]

Other accounts don't make me feel safe, however.

Nahum Ward visited the saltpeter-mining operation in Mammoth Cave about 1815, a few years after the New Madrid earthquake and aftershocks that occurred during the winter of 1811–1812. He wrote, "One of my guides informed me he was at the *second hoppers* [the vats used to leach saltpeter from sediment] in 1812, with several workmen, when those heavy shocks came on, which were so severely felt in this country. He said that about five minutes before the shock, a heavy rumbling noise was heard, coming out of the cave like a mighty wind, when that ceased the rocks cracked, and all appeared to be going in a moment to final destruction. However no one was injured, alth'o large rocks fell in some part of the cave."[8]

Elisa Kagan from the Institute of Earth Sciences at the Hebrew University of Jerusalem interviewed two employees of Soreq Cave, a show cave in Israel, who were in the cave during a 5.1-magnitude earthquake on February 11, 2004.

> Tamir Cohen was leaning against the metal railing of the main entrance viewing area when he felt the shaking begin. The railing began to move, "as if someone had banged on it" and caused it to vibrate. He described it as if the "railing was twisting." He heard low noises of objects rolling in the cave (there are many loose rocks and loose broken speleothems on the cave floor). The people sitting on the concrete bench also felt the shaking, Tamir says.
>
> Sylvie was also on the viewing platform during the earthquake. She felt as if the whole cave was shaking, but she says it [might have] be[en] only the concrete platform. She heard noises, such as rocks rolling around. A "rock" fell from the ceiling. No one (as of yet) has been able to show us the piece that fell. She says they looked for other changes to the cave and found none. Sylvie says that after comparison with witnesses on the surface above the cave, she felt the earthquake at least as strong as those above ground.[9]

The last two accounts are less scary if you keep in mind that the guides at Soreq Cave were near the entrance, a place more susceptible to earthquakes.

The New Madrid Earthquake

Caves are earthquake resistant but not earthquake proof. The New Madrid earthquake and its aftershocks between December 16, 1811, and March 15, 1812—centered around the Mississippi River town New Madrid on the border between Kentucky and what would eventually be Missouri—shook up the Mammoth Cave saltpeter-mining operation. This was the only earthquake ever felt in the cave (although in 1987 a rock estimated to weigh 1,200 pounds fell in a passage called Audubon Avenue during a 3.2 earthquake with the epicenter near Erlanger, Kentucky. No one was in the cave during the earthquake, but some people felt it on the surface).

The magnitude of the New Madrid earthquake was the main reason saltpeter miners felt it underground, but the piles of spent dirt from the mining operation may have contributed to the Rotunda's shakability (I made up that word, but it works) because sediment moves more easily than bedrock.

The New Madrid earthquake's strongest shocks on December 16, 1811, and February 7, 1812, are often estimated to have been 7.7 on the Richter scale,[10] but some sources estimate they were as high as 8.3,[11] bumping this earthquake from the category of "major" to "great." Seismographs didn't exist back then, so these magnitude estimates are based on journal entries, newspaper articles, and other firsthand accounts.

If the New Madrid earthquake was 8 or higher, it ranks as the biggest earthquake in the recorded history of the lower forty-eight states (it is outranked by several quakes in Alaska in the twentieth century and one on the Washington coast in 1700 that left physical evidence but no written records in the United States). If the magnitude was a mere 7.7, that knocks this earth-

How Big Is Big?

Earthquakes are often measured by the Richter scale from 1.0 up: there's no upper limit! The 3.2 earthquake that caused the rockfall in Audubon Avenue was in the minor category. An earthquake between 7.0 and 7.9 is considered major, and a quake that is an 8.0 and higher is great. Each higher number—for instance, from 7.0 to 8.0—represents ten times as strong an earthquake, so a quake that rates 8 on the Richter scale is ten times stronger than one that rates 7.0.

quake down to a ranking of number 18 in size for the whole United States.[12]

Whichever measurement is used, though, the New Madrid quake was big. In addition to three main shocks, there were about 2,000 aftershocks of various magnitudes during the winter of 1811–1812.[13] People felt movement as far away as Montreal, Canada; New Hampshire; and the Atlantic coast of Florida.[14] As many as 1,500 people, mostly American Indians, may have died, though some estimates of mortality are as low as 11.[15]

Along with the real damage came exaggeration. Stories say the quake rang church bells in Boston, but, according to earthquake expert Otto Nuttli, there are no records of any such thing happening. People in Massachusetts felt some movement, and a Boston newspaper said a church bell rang in Pennsylvania. The New Madrid earthquake had the power to cause buildings closer to the New Madrid Seismic Zone, including church steeples, to sway and collapse, but not as far away as New England.

Stories say the Mississippi River ran backward, which it couldn't literally do. But the Reelfoot Scarp did thrust up, which could force the water back for a short time, making it look as if the river ran backward.[16]

The New Madrid Seismic Zone and Intraplate Earthquakes

More than 90 percent of all earthquakes occur near tectonic plate boundaries, where the earth's plates either move apart, slide past one another, or override and push each other down. That's why many earthquakes happen on the West Coast, where the North American plate meets the Pacific plate and the small Juan de Fuca plate. These earthquakes are interplate quakes.

The New Madrid earthquake occurred in the middle of the North American plate, far from the plate's boundaries with the Pacific plate on the West Coast and the Eurasian plate in the middle of the Atlantic Ocean. The New Madrid earthquake was an intraplate quake.

Earthquake experts Arch Johnston and Eugene Schweig use a famous quote from Winston Churchill to describe the New Madrid earthquake: "it is a riddle wrapped in a mystery inside an enigma" (though Churchill was talking about Russia, not earthquakes).[17]

Intraplate quakes are mysterious enigmas, so we don't know much about them, but here's a little info. Big intraplate earthquakes tend to be less common and have more time in between quakes than interplate earthquakes. But intraplate quakes release more force, and the stronger rocks in the middle of plates cause seismic waves to dissipate more slowly, so they travel farther and more efficiently than the energy going over weaker rocks near plate boundaries.[18] The North American plate isn't completely solid—it has cracks and imperfections. The New Madrid Seismic Zone sits on the Reelfoot Rift, a fault caused when the earth's crust was pulling apart there roughly 600 million years ago. Today, plates to the east and west are compressing the rift,[19] causing small earthquakes almost daily, but we don't feel them.[20] No one knows exactly when the next big one will happen, but seismic studies show big earthquakes occur in the New Madrid area about every 500 years.[21]

These rocks could no longer resist gravity.

Moral of the Story

Rocks don't often fall in Mammoth Cave, but they sometimes do. Earthquakes don't tend to affect caves, but they can. When it comes to rockfall and earthquakes, never say never.

Special thanks to Rickard A. Olson, ecologist at Mammoth Cave National Park, and Rickard S. Toomey III, director at the Mammoth Cave International Center for Science and Learning, for their geological expertise and advice.

3

Bones

What's Left of Exciting but Dead Animals in Mammoth Cave

On tours, cave visitors may see marine fossils such as coral and crinoids that became fossilized as the limestone formed under the sea, but the cave also holds treasures that few people see—BONES! Pieces of the remains of a mastodon, a giant short-faced bear, a saber-toothed cat, a vampire bat, a horse—bones of exciting mammals galore.

Most of them are surface-animal bones, so they are allochthonous (uh-LOK-thuh-nuhs) fossils. *Allo-* means "other," so *allochthonous* means the bones originated in a place other than where we found them.

If the bones originated elsewhere, how did they get in the cave? Most of the critters either

- Fell down a pit, wandered in and got lost, or met some similar misfortune or
- Died above ground and got carried in by a predator, a human, a woodrat or other bone-collecting critter or
- Died above ground and got washed down a sinkhole, entrance, or sinking stream or
- Lived in the cave part-time, such as bats or woodrats.

Whose Bones Are in the Cave?

Mammoth, No—Mastodon, Yes!

In answering the question "Were there any mammoths in Mammoth Cave?" cave guides for years said, "No." In 1979, however, the answer became "maybe." Now the answer is, "Mammoths, no, but mastodons, yes."

In 1979, Ronald Wilson, a curatorial assistant at the Carnegie Museum of Natural History in Pittsburgh, joined a team of Cave Research Foundation cavers to get a giant short-faced bear jawbone from the Proctor Cave section of the Mammoth Cave system. Getting to the jawbone included a tough canyon descent that was beyond Wilson's skill, so he waited in Frost Avenue for the rest of the party to get the jaw. While poking around, Wilson found several bone fragments near a sinkhole where water trickles in from the surface. He returned to the site later to look again and found more bones and part of a tusk.[1] No mammoth or mastodon visited Frost Avenue (you have to belly crawl); rather, water washed the bones and tusks down the sinkhole.

The tusk fragments went to the Carnegie Museum, where they were labeled as proboscidean, or of the order Proboscidea, which includes mammoths, mastodons, and modern elephants.[2] No one knew whether the tusks were mammoth or mastodon. Either was possible; paleontologists had found mastodon and mammoth bones in other Kentucky caves.[3] (Paleontologists ruled out modern elephant due to the lack of elephants in Kentucky.)

An ivory expert can tell what animal a tusk came from by the angles at which the layers of ivory intersect; mastodon ivory averages a 125-degree angle, and mammoth ivory about an 87-degree angle. In the late 1990s, I tried to get the tusks from Carnegie Museum to a proboscidean expert to identify, but I was thwarted by all the channels I needed to take to get permission. I secretly hoped for a mammoth.

In 2001, paleontologists Rick Toomey and Mona Colburn and park ecologist Rick Olson found pieces of tusk and fifteen small bone fragments (probably pieces of ribs and limbs) in Frost Avenue. Colburn took the tusk pieces to the Illinois State Museum in Springfield, cleaned them, and sent them to mastodon authority Dr. Dan Fisher at the University of Michigan. Dr. Fisher measured 120-degree angles in the tusk pieces—indicating that it came from a mastodon. The other bone fragments found near the tusk fragments are probably mastodon, too.[4] Darn.

What's the difference?

The most obvious differences between these prehistoric proboscideans were their size and diets. Mastodons (genus *Mammut*) stood about eight to ten feet tall, mammoths (genus *Mammuthus*) about seven to fourteen feet. Mastodons' teeth indicate they were browsers that ate leaves and twigs. Mammoths had teeth suited for grazing on grasses.

Giant Short-faced Bear

So what about the jawbone and the bear it came from?

Giant short-faced bears (*Arctodus simus*) are well named—they weighed 1,800 pounds or more and stood about eleven feet tall when upright. For comparison, a polar bear, the largest liv-

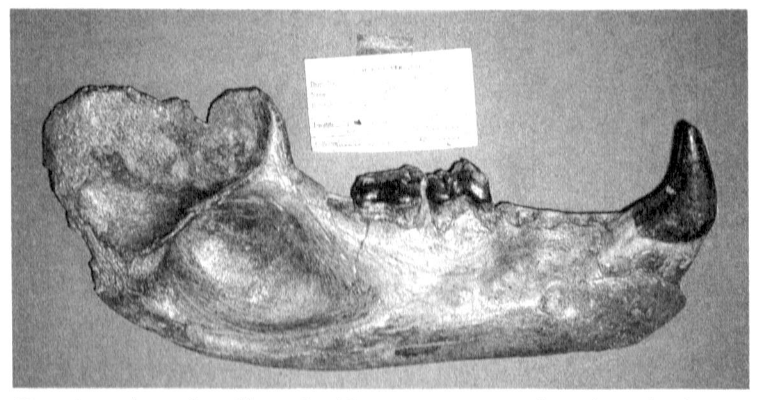

The giant short-faced bear had huge jaws to match its huge body.

ing bear, weighs up to 1,500 pounds and stands less than ten feet upright. New studies indicate that giant short-faced bears, though once thought to be strict carnivores, were omnivores, like most modern bears.

Like the mastodon bones, the bear's jawbone washed into the cave.

More Bears—a Tale of Two Tibias

Ron Wilson found another bear bone in Sophie's Avenue in 1978, the tibia of a black bear (*Ursus americanus*)—a lower back leg bone. He took it to the Carnegie Museum.

In 2001, Rick Olson, on a paleontological research trip with Rick Toomey and Mona Colburn found a black bear's left tibia in Sophy's Avenue. Wilson listed the bone he found as the *right* tibia, so the group assumed the new bone belonged to the same bear for the bone Wilson found. Colburn looked at an old photo of the bear tibia that Wilson found and thought it looked like the left tibia, not the right. She went to the Carnegie Museum for a better look. The first bone was also a left tibia—the two bones came from two bears!

A "Small" Cat

The jaw and two leg bones—the right tibia and right fibula—of a bobcat (*Lynx rufus*) were found in a stream in the Proctor section of Mammoth Cave. The fibula was covered with redeposited calcium carbonate, indicating the bone had been there a long time.[5]

A Big Cat—Really Big

Mona Colburn found another cat bone in Proctor. This leg bone, a fibula, came from no kitty cat. Colburn compared the bone to the bones of modern cougar, jaguar, and lion—pipsqueaks compared to this cat. She believed it was from the extinct saber-toothed cat (*Smilodon fatalis*).

To make sure this surprising find really was a saber-toothed cat, Colburn compared it to saber-toothed cat bones at the Illinois State Museum in Springfield, the Field Museum in Chicago, and the Smithsonian National Museum of Natural History in Washington, D.C. Colburn's colleague from the Illinois State Museum took the bone to experts at the University of California, Los Angeles, and the Los Angeles County Museum, which has the fossils from Rancho La Brea Tar Pits, the world's biggest collection of saber-toothed cat bones. They all agreed it was a saber-toothed cat. So far it's the only one found in Kentucky.

The saber-toothed cat was about six and a half feet long and about three feet tall at the shoulder, and it weighed about 765 to 974 pounds.[6] For comparison, a big cougar weighs about 220 pounds, and a Siberian tiger, the biggest cat alive, weighs up to 660 pounds.

Fox Avenue or Raccoon Avenue?

In 1929, George Morrison, the owner of Mammoth Cave's New Entrance in pre–national park days, wrote that Fox Avenue was "named due to the finding of the remains of three foxes at the extreme end of the terminal of this avenue."[7]

Morrison apparently didn't know bones. In 1975, Cave Research Foundation explorer Diana Daunt said she had found one skeleton in Fox Avenue—that of a raccoon. Twenty-five years later, cave guide Chuck DeCroix led paleontologists Rick Toomey and Mona Colburn to look at the Fox Avenue raccoon. In addition to the skull and bones Daunt mentioned, they found a second raccoon skull. These raccoons may be two of Morrison's "foxes."[8]

Frozen Niagara Treasures

In 2000, Rick Toomey and Mona Colburn did some paleontological research at Mammoth Cave's Frozen Niagara Entrance. Though the entrance as we know it was blasted open in 1924, an opening that allowed a wide assortment of animal remains to wash in existed before that. The researchers found frog, salamander, turtle, snake, lizard, bird, deer, and raccoon bones—interesting finds, but they also found animals even less expected.[9]

A Horse? Of Course!

A tooth fragment of a prehistoric horse had washed in. Extinct for 10,000 to 12,000 years, prehistoric horses native to North America had teeth similar to modern horses—they were both from the genus *Equus*. Another ancient horse's tooth and skull fragment were found in the Proctor Cave section of Mammoth Cave.[10]

Pocket Gopher

The paleontologists also found pocket gopher teeth. The gopher was either a plains pocket gopher (*Geomys bursarius*), which today lives in western and midwestern states, or a southeastern pocket gopher (*Geomys pinetis*), which lives in Florida, Alabama, and Georgia. Kentucky is free of pocket gophers today, but these teeth and other fossils found in Kentucky show they lived here in the past.[11]

Pocket gophers got their name from the furry pockets on the outside of their cheeks that they use for carrying food. Their lips close behind their teeth, so as they gnaw through dirt and roots, they don't get dirt in their mouths—a handy trait for their burrowing lifestyle.[12]

Pocket gophers live in dug-out burrows, not in caves, so the remains found at the Frozen Niagara Entrance probably washed into the cave.

A Beautiful Armadillo

The paleontologists at Frozen Niagara Entrance also found some armor plates from an extinct beautiful armadillo (*Dasy-*

pus bellus). The beautiful armadillo was similar to the modern nine-banded armadillo but larger.[13]

Peccary

The tooth of a flat-headed peccary (*Platygonus compressus*) was among the many teeth found at Frozen Niagara Entrance. These prehistoric herd animals liked cave entrances; peccary remains were also found in the Proctor Cave section of Mammoth and many other Kentucky caves. The large number of peccary fossils in North America indicates they were "probably the most numerous medium-sized mammal during the Pleistocene" (about 1.8 million to 10,000 years ago) according *Pleistocene Mammals of North America.*[14]

Peccaries (also called javelinas) are related to Old World pigs but are not true pigs. In case you meet a modern piglike animal and are confused, you can tell the difference by their tusks; peccaries' tusks point down; true pigs' tusks curve up.

Water Rat

Part of the jaw of an extinct water rat (*Neofiber leonardi*) was among the Frozen Niagara bones. This species' modern relative, the Florida water rat, also known as the round-tailed muskrat, lives in Florida and southeastern Georgia.[15]

Other Parts of the Cave

A Marten, Gone from Kentucky but Not Forgotten

Park ecologist Rick Olson found the skull of a marten (*Martes americana*) in Proctor Trunk. Researchers later found other marten vertebrae, pieces of a jaw, a front leg, and foot bones. A researcher named Kenneth Dearolf also found marten bones in Fossil Avenue in the New Discovery section of the cave in 1941.[16]

Also called the American marten, pine marten, and American sable, martens no longer live in Kentucky. They live in Canada, some northern states, and at high elevations—climates different from that of modern Kentucky. During the late Pleistocene (10,000–20,000 years ago), the colder climate allowed martens to live as far south as Alabama.[17]

Elk

In the early 1980s, Rick Olson found an elk (*Cervus canadensis*) tooth in a rimstone pool by Logsdon River in the Proctor Cave section. Elk had been overhunted and eliminated from Kentucky by the mid-1800s but were reintroduced into eastern Kentucky in 1997.

Tapir

Cave Research Foundation explorers found a tapir (*Tapirus veroensis*) tooth in Proctor Crawl of the Proctor Cave section of Mammoth Cave in 1978.[18] This extinct tapir has relatives in Central and South America and Southeast Asia. Tapirs, both extinct and modern, look like pigs with trunks, but they are no relation to pigs or elephants. They are instead closer to horses and rhinos.

Vampire Bat

Finding a bat femur in the cave would normally seem unexciting, but Rick Toomey and Mona Colburn found a femur with an unusual shape in Backsliders' Alley. The femur was built for walking and jumping, something most bats don't do—except vampire bats (genus *Desmodus*), that is.

A vampire bat lands near its victim, walks up to it on its legs and wings, slices the victim's skin, and spends about thirty minutes lapping up the blood. The victims are usually unharmed and often unaware of their role as meal provider, but the wound can cause infection.

The only mammals that eat nothing but blood, modern

vampire bats live in southern Mexico, Central America, and South America; they like the warm winters in these areas. If we assume that the extinct bat (*Desmodus stocki*) at Mammoth Cave had a similar taste for warm weather (which we don't know), we can surmise that the bone dates back to when Kentucky had a warmer climate. Or this bat may have been a long way from home, considering that this femur is the only vampire bat bone found in Kentucky.[19]

Long-Gone Bats

Free-tailed bat (genus *Tadarida)* bones are scattered from near the Cataracts to Mayme's Stoop sections of the cave. The Mexican free-tail (*Tadarida brasilienses*) is the only living American species of the genus *Tadarida,* but the bats in Mammoth Cave could be from the extinct species Constantine's bat (*Tadarida contantinei*). On the trail in Chief City, you can see chocolate-colored guano that is common to free-tailed bats.

Free-tailed bats live in climates warmer than that of present-day Kentucky. The Sangamon Interglacial period 75,000 to 130,000 years ago is the most recent period that free-tails would have been comfy at Mammoth Cave, but they may have lived here during earlier interglacial periods.[20]

Hellbender

While researching hellbenders, I learned that there are two movies titled *Hellbenders*—a horror flick and a Western. There is also a *Hellbenders* video game, which involves shooting ruthless killing machines before they kill you. A critter named "hellbender" sounds scary and wild, worthy of movies and video games. Little did the movie and game people know that this animal is just a harmless (but big) salamander.

Paleontologists found chalky fragments of hellbender bones in a passage called Backslider's Alley. They could identify only the bones' genus, *Cryptobranchus,* but not the species.

Not cave dwellers, hellbenders live in the Green River. Aquatic, they breathe underwater but have lungs instead of gills: they breathe through their skin!

4

Secret Lives of Cave Critters Revealed!

Writers like bats. There are lots of bat books. Bats even show up in fiction—*Dracula* is one famous example. But how many novels have you read about cave crawdads? None. Even nonfiction books about cave animals other than bats are scarce. It's time to bring cave critters out of the dark (figuratively speaking).

Roughly 130 species regularly live in Mammoth Cave. If you include animals that occasionally wander or fall in, the number goes up.[1] That's a lot of species for a cave. But compared to the surface world, with its sunlight and plants that make life easy, a cave is a tough place to live.

Cave-Critter Lingo

Troglobite	A permanent cave dweller. Some biologists use the term specifically for land animals, but it can include aquatic critters.
Troglobiont	The same thing as a troglobite. This term is popular with cave biologists.
Troglophile	An animal that may live its entire life in a cave but can also live above ground or in a cavelike habitat. The term literally means "cave lover."
Trogloxene	A cave dweller that spends part of its life in a cave but must go to the surface (at least

	occasionally) to feed. The term means "cave visitor."
Accidental	A surface animal that accidentally gets into a cave.

Cave Food

The absence of light and plants beyond cave entrances makes for just these few dining options.

Eating out. Trogloxenes, such as bats, wood rats, and cave crickets, leave the cave at night to eat at the surface smorgasbord of plants, fungi, prey, and carrion.

Predation. Spiders, fish, crayfish, and other predators eat other cave animals.

Delivery by water. Water filters through leaves and other organic matter on the surface, washing nutrients into the cave through seeps, sinking streams, and sinkholes. It's a process similar to making tea.

Delivery by animal. Animals that come and go from the cave provide food for troglobites and troglophiles, which never leave the cave. The food may be the animals' dead bodies, eggs, or guano (a.k.a. poop). Crickets provide most of the guano eaten in Mammoth Cave, though critters also eat guano from cave beetles, bats, and woodrats.

Aquatic Residents

Cave Fish

Two species of troglobitic fish live in Mammoth Cave. The five-inch-long northern cave fish (*Amblyopsis spelaea*) lives in the generally calm waters of base-level rivers. The three-inch-long southern cave fish (*Typhlicthys subteraneus*) tends to live in faster-moving streams above the water table. The northern cave fish can live to the old age of seventy years,[2] and the southern cave fish lives about fifteen years; surface fish in the same

A northern cave fish in Mammoth Cave's Mystic River.

family live only a year or two.[3] Cave fish eat amphipods, copepods, isopods (types of crustaceans), crayfish, and, when times get tough, each other!

Eyeless or blind? The fish in Mammoth Cave have historically and most commonly been called "eyeless fish," but Dr. Tom Poulson, a biologist who has done a great deal of research at Mammoth Cave, prefers the term *blind fish* because even though these fish don't have working eyeballs, they do have some remnants of eyes.[4] Both names are acceptable; after all, prairie dogs aren't really dogs, and guinea pigs aren't really pigs.

The benefit of no eyes. Total darkness is totally dark, no matter how good your eyesight is. Scientists have long wondered how having no eyes benefits eyeless fish and other troglobites. Biologists have some ideas.

A gene that regulates metabolism (how efficiently the body uses energy) may be linked to a gene that affects eye development. To a surface animal, having better metabolism but no eyes would be a bigger loss than a gain, but lack of vision is no loss for a cave fish.

Another possible benefit of the absence of eyes is that it leaves more room in the skull for olfactory receptors, giving the fish a better sense of smell. To most animals, the loss of vision would outweigh the gain of a better nose, but to a cave fish this loss leads to a big gain.[5]

Lateral line system. Along with their sense of smell, eyeless fish use their sense of touch to navigate dark waters. Like other fish and some amphibians, cave fish have a *lateral line system* that runs along their sides, snouts, and jaws. The lateral line system feels differences in water pressure caused by other creatures moving or waves bouncing off of an obstacle as the fish swims toward it. This is why fish in aquariums don't swim into the glass.

The lateral line system has sensory cells with hairs covered with gelatinous caps called neuromasts. The neuromasts move and the hairs bend as pressure waves hit them.

The lateral line system can also detect low-frequency sounds because sound waves are pressure waves.[6]

The cave fish's long pectoral and tail fins help them swim more efficiently in two ways; the longer glide they get with each thrust conserves energy (important in a food-poor environment) and cuts back on creating waves that interfere with the neuromast receptors on their lateral line systems.

The two cave fish's different lifestyles call for different lateral line systems. The northern cave fish has a neuromast more sensitive to general movement, which suits its active lifestyle and hunting of larger prey in slower base-level streams. The southern cave fish is more sensitive to the direction the movement is coming from, an advantage to these less-active fish that eat smaller prey (mostly copepods) in faster water. Both species have a larger lateral line system with bigger and more neuromast cells than their surface relatives.[7]

Spring Fish

As the name implies, the spring fish (*Forbesichthys agassizi*) lives near springs. If a spring fish gets washed into the cave, it can survive and even reproduce if enough food washes in. Spring fish have partially reduced eyes, a compromise for life in and out of the cave. When the going gets tough (i.e., not enough food), the spring fish may starve to death in the cave.[8]

Unlucky Surface Fish

In addition to washing food into the cave for aquatic cave critters, back flooding also sometimes washes in surface fish, including bullhead and madtom catfish, blue gills, minnows, gar, and carp. They won't live long. Their misfortune, however, is good for cave animals—the dead fish become lunch.

In spite of their inability to survive long in the cave, surface fish can begin to look like cave fish while they do survive. They turn pale in the dark because their pigment will cluster in the center of cells instead of spreading out.[9] The fishes' chromatophores (cells that produce pigment) eventually stop producing pigment.[10] That's why the trout introduced and fed at the Lost Sea, a show cave in Tennessee, are so pale; they have lost their suntans!

Sculpin

The banded sculpin (*Cottus carolinae*), a surface fish, is often found in base-level streams in Mammoth Cave. They're weird (my term, not official) because they fare better than other surface fish in the cave and may even live their whole lives there. The sculpin in Mammoth Cave look the same as sculpin on the surface. Sculpin are broad in the head and graze on insect larvae in the stream bottom in rocky habitats in surface streams.

In the cave, they may eat biofilm that consists of bacteria and fungi, but we know little about their cave lifestyle.

Crayfish

The pale, eyeless cave crayfish (*Orconectes pellucidus*) live in water but can travel over dry land from one body of water to another. This ability makes it possible for cave crayfish to be in any aquatic community in the cave, from rimstone pools at upper levels to the lowest base-level streams.

Compared to their surface cousins, cave crayfish have long antennae with extra receptors to cope in the dark. One antenna feels objects and movement; the other detects smells.[11] Minus antennae, the crayfish is about one to three inches long.

A study on a related crayfish (*Orconectes australis*) indicates that the crayfish react to light but cannot tell what direction it comes from.[12] Several species of crayfish have photoreceptors near their tails that detect light.[13] No studies have been published on *Orconectes pellucidus*'s reaction to light, but it probably is wired like its relatives, and people have seen this crayfish turn away from light.[14]

Cave crayfish may live more than sixty years; local surface crayfish tend to live about two to two and a half years.[15] Many cave critters, including eyeless fish, bats, and cave crickets, live longer than their surface relatives. Because they don't get as much to eat as surface animals, they can't afford to burn as much energy. Making babies takes energy, so cave animals have fewer offspring at a time and may not reproduce every year. A long life gives them time to reproduce at longer intervals.

Cavers sometimes see the crayfish species *Cambarus tenebrosus* (its common name is just "crayfish" or "crawdad"). This six-inch crayfish usually lives outside the cave, has eyes, and is stockier and darker than the cave crayfish. *Cambarus* can survive and reproduce in the cave if enough nutrients wash in for it to eat. In polluted water, it can even outcompete the cave cray-

(Above) A cave crayfish with eggs. *(Below)* This surface crayfish got washed into River Styx, named for the river in the underworld of Greek mythology. How fitting for a doomed crayfish.

The endangered Kentucky cave shrimp.

fish.[16] But if times get tough, meaning there's little rain to wash in food, this crayfish won't survive in the cave.

Shrimp

The endangered Kentucky cave shrimp lives only in the Mammoth Cave area in base-level streams and pools. Discovered in Roaring River in 1901, *Palaemonias ganteri* was named after cave manager H. C. Ganter.[17]

In the 1970s, people feared this shrimp was extinct—no one had seen any for years. On September 1, 1979, however, a scientist found one dead specimen in the Roaring River Shrimp Pools. Sixteen months later, someone saw three live shrimp in the Flint Ridge section of the cave.[18] Today, when biologists look for cave shrimp, they may count as many as seventy in one day![19] This species' recovery is probably due to cleaner water.

These nearly clear, inch-long shrimp graze on biofilm made of bacteria and protozoa.[20]

Isopods and Amphipods

Tiny aquatic isopods live in cave streams. These crustaceans look like grains of rice with legs—white, flat from top to bot-

An isopod in a pool called Devil's Cooling Tub.

tom, and about half an inch long. Mammoth Cave has two species of isopods; they look similar but have different habitats. *Caecidotea stygius,* the isopod you may see in wet areas on crawling tours, lives in relative safety in upper-level streams with few predators. *Caecidotea bicrenata whitei* lives in rivers, where it becomes lunch for fish and crayfish.[21] Isopods eat bacteria, protozoa, and nutrients from organic matter that washes into the cave.

This amphipod may become lunch for a fish or crayfish.

The amphipod, another tiny white crustacean, is flat from side to side and has a curled tail. Mammoth has two species of blind troglobitic amphipods (they can't live outside the cave) and one species of troglophile amphipod (it can live outside the cave) that can see and lives near entrances that have more food.[22]

Terrestrial Mammoth Cave Residents

Salamanders

Reddish orange with dark spots, the cave salamander (*Eurycea lucifuga*) begins life in the water. The larva goes through metamorphosis, and the adult salamander becomes terrestrial. These salamanders live in cave entrances, where they eat mostly cave crickets. They also live in other dark, damp areas, such as under logs and rocks, where they eat insects, mites, ticks, isopods, and worms. The light-colored, short-tailed young darken and grow longer tails as they age. Both males and females are about four to eight inches long, but the male has a more swollen snout.[23]

Cave salamanders like cave entrances with plenty of crickets to eat.

The well-named slimy salamander.

The black, lightly spotted slimy salamander (*Plethodon glutinosus*) also lives in and near the cave. Slimy salamanders get their name from a slimy liquid that secretes through their skin. Don't pick one up: you may injure it, and the slime is hard to wash off! Slimy salamanders lay their eggs in damp areas, but they are totally terrestrial. They live under rocks and logs and in ravines as well as in caves. Slimy salamanders eat ants, beetles, sowbugs, and worms.

Cave Crickets

Mammoth Cave visitors see more cave crickets (*Hadenoecus subterraneus*) than any other cave critter; when the lights are on, the crickets can see, too. In the dark, they rely on their long front antennae to touch and smell their way around. Small antennae called "cerci" on their hind ends feel breezes.

Cave crickets can go deep in the cave as long as it's damp, but they tend to roost within about three hundred feet of entrances. Females go deeper in the cave to lay eggs.[24]

Crickets can eat up to twice their body weight in one meal,

The cave cricket, an important member of the cave community.

so they leave the cave only to eat about every eleven to thirteen days. They'll eat just about anything soft and stinky—mushrooms, dead insects, feces, berries, flowers, and even other crickets. But if the cricket now labeled as food is injured, the predatory cricket won't eat the victim's crop (part of the digestive system), maybe to avoid getting sick from the same thing that may have made the victim sick in the first place.[25]

A thin cuticle (the outer layer of the exoskeleton) makes cave crickets sensitive to moisture loss and extreme temperatures, hot or cold. More crickets come out of the cave when it's rainy because they will lose less water, and the dampness makes food stinkier and easier to find. When the temperature drops into the low 40s Fahrenheit, the crickets wait for a warmer night to feed. They can't take heat either, though; they lose water in temperatures above the high 60s and will soon die in temperatures above the high 70s.[26]

High temperature also affects microbes (such as bacteria and yeast) that live in the crickets' crops. The microbes help them digest, but when it's hot, the microbes produce gas or toxic metabolites that cause the crop to expand and sometimes rupture, killing the cricket.[27]

Cave crickets are like pizza-delivery service for troglobites.[28] Springtails, bristletails, millipedes, snails, and other troglobites eat cricket guano (it may not be pizza, but cave critters can't be choosey).

Crickets supply food to cave animals in other ways as well. The critters that eat guano get eaten by pseudoscorpions and predatory mites. A species of cave beetle (*Neaphaenops tellkampfi*) eats cricket eggs (more on this beetle later), and the orb weaver spider (*Meta americana*) preys on crickets that get caught in the spiders' webs.

Cave crickets live four or more years,[29] a long time for an insect. They need a long life to reproduce enough offspring for the species to survive. A single cave cricket lays about 167 eggs a year, many of which get eaten. By comparison, the shorter-lived common house cricket lays about 728 eggs a year, and the American cockroach lays as many as 1,000![30]

The female cricket lays its eggs in holes it pokes in sandy sediment. But it will lay the eggs only if the moisture of the sediment is suitable; if it's not, the cricket pokes more holes until it finds the right moisture. Once it lays the eggs, it goes on its way, leaving the eggs to fend for themselves.

Camel Crickets

Mammoth Cave has two species of camel crickets (*Ceuthophilus stygius* and *Ceuthophilus thomasi*).

Less suited for long-term cave life than cave crickets, camel crickets go outside to eat every two or three days, so they live near

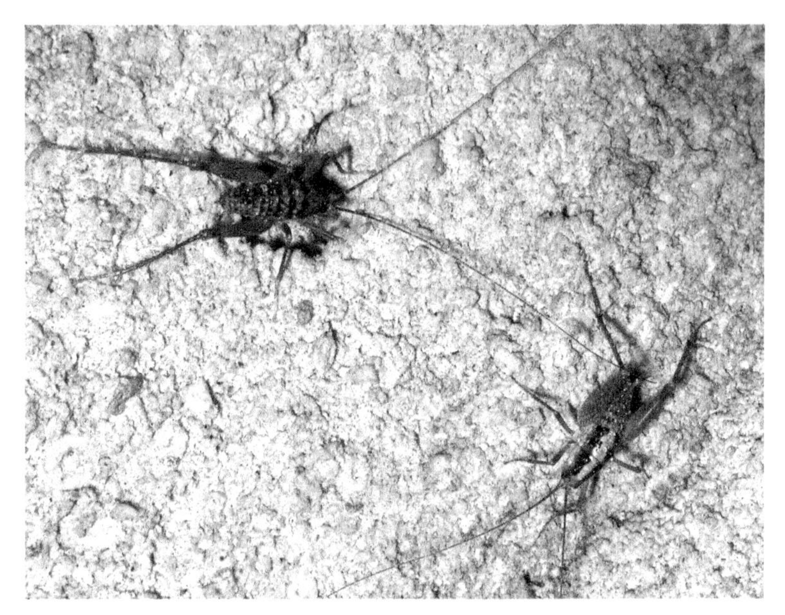

These camel crickets are two different species; notice the different stripes.

damp entrances. Because of their frequent trips out, they have to be able to endure a wide range of weather and so have a thicker cuticle than cave crickets and can handle dry conditions better.

Because camel crickets go out to eat more often than cave crickets, they eat less on each foraging trip, about 34 to 39 percent of their body weight.[31] They also lack the longevity of cave crickets, living for only about a year.[32]

Beetles

Life for the troglobitic beetle (*Neaphaenops tellkampfi*) is like Easter and Thanksgiving—hunting for eggs, then eating a big meal. The beetles use about 75 percent of their energy but only 5 percent of their time hunting for cricket eggs. A beetle can afford to relax after working hard to find food because one egg is almost as big as the beetle, and it takes the beetle about fifty days to burn off the weight gained.

A cricket egg makes a big meal for this beetle.

A study in Great Onyx Cave (in the park but separate from Mammoth Cave) showed these beetles ate more than 85 percent of the cricket eggs laid.[33] Great Onyx has a big beetle population, probably causing a higher predation rate than in other caves.[34] No other critter eats cricket eggs, and during egg season this beetle eats little else. When eggs are unavailable, it preys on whatever it can get.[35]

The egg-eating *Neaphaenops* is just one of many beetles. Mammoth Cave (like the world in general) has more types of beetles than of any other animal. Other beetle species eat dead animals, springtails, worms, and other beetles.[36]

Other Insects

Troglobitic bristletails have bristly antennae called cerci on their hind ends—hence, the name.

Caves tend not to have many booklice because caves lack

books, but Mammoth Cave has a species of lice in the back of Audubon Avenue,[37] Onyx Colonnade, and a few other passages.

Several types of flies, including one troglobite, live in the cave. Near Frozen Niagara Entrance, you can spot four species called heleomyzids[38] that look like houseflies. Though they are common, we don't know what they eat, how long they live, or why the cave suits them.

Buglike Things That Aren't Insects

Springtails (*Folsomia candida*) used to be considered insects but have been kicked out of the exclusive club of class Insecta. As their name implies, springtails can escape from predators by springing away with a leverlike body part called a "furcula." Several types of springtails live in the cave; they range from $^{1}/_{10}$ inch to $^{3}/_{10}$ inch long, so they are hard to spot. They eat cricket guano, leaf litter, and woodrat waste.[39]

Springtails in a cave pool.

Mites, related to ticks, are food for other small cave critters. A predatory mite preys on other mites.

Pseudoscorpions (possibly *Kleptochthoius cerberus*) look like tiny scorpions, but they aren't—the prefix *pseudo-* means "false." These arachnids often live in woodrat latrines or near cricket guano communities, and they eat mites and springtails.

You have to get down in the dirt and look close to spot a pseudoscorpion.

Two types of troglobitic millipedes live in the cave. The more common one, *Scoterpes copei,* is up to an inch long and hairy, and it lives up to twelve years. The hairs have a liquid on them, maybe to protect the millipede from predators. The term *millipede* means "1,000 feet," but most members of this species have only between 34 and 400 feet.

A cave millipede.

A species of centipede lives in the cave near entrances.[40] The term *centipede* means "100 feet." Some species live up to the name; most centipedes have between 20 and 300 feet.

Spiders and Harvestmen

In the cave scene in *Raiders of the Lost Ark,* the cave is crawling with tarantulas—that's how some arachnophobic people imagine caves. But for the most part spiders don't like caves, though Mammoth has a few. Species of wolf and fishing spiders, which look like tarantulas, occasionally hang around entrances. They look scary, but if you leave them alone, they will leave you alone.

Orb weaver spiders (*Meta americana*) live in the cave's Frozen Niagara area. The big cricket population there gives spiders plenty to eat. The cave also has two seldom-seen troglobitic spiders.

The cave orb weaver spider waits for a cave cricket to get caught in its web.

The cave harvestman is a daddy longlegs, not a spider.

Troglobitic harvestmen (daddy longlegs, *Phalangodes armata*) live in the cave.[41] They eat tiny invertebrates. Though people often confuse harvestmen with spiders, they are a different type of arachnid. Harvestmen don't spin webs, and, contrary to popular belief, their venom is not the most potent in the world because they don't have any!

Woodrats

People have a love-hate relationship with rodents. We see them as disgusting yet cute. The Allegheny woodrat (*Neotoma magister*) is on the cute side.

Woodrats are an important part of the cave's food-delivery service; they go above ground to eat fruit, nuts, seeds, leaves, and fungi and bring some of this food back to the cave. This behavior makes them good indicators of how the ecosystems

Woodrats visit the woods but live in the cave.

above and below ground are doing because changes on the surface impact their behavior.[42]

The Allegheny woodrat is listed as endangered or threatened in more states than any other rodent.[43] Deforestation and parasitism by the raccoon roundworm—which the woodrats can catch when they carry raccoon poop and other scat back to their nests—may contribute to their decline.[44]

Both sexes of woodrats have glands on their chests that produce strong-smelling oil,[45] which they use to communicate, but we don't know in what way. The extra-strong smell among breeding males implies that the scent is used to attract a mate. The oil may also say "keep off my turf"[46] and help woodrats navigate the cave. Their long whiskers also help them feel their way through the dark.[47]

Woodrats often build nests near the Frozen Niagara Entrance. This site seems to be prime real estate, if the number of nests found there is any indication. But the same woodrats are not rebuilding every year—this species tends to live only one or two years in the wild (though as long as six years in captivity), so the kids may be inheriting the real estate.[48]

Around their nest of dry grass, bark, and plants, woodrats have a midden of debris. In wild parts of the cave, middens consist of leaves, nut hulls, and other natural things that woodrats drag in. Along tour trails, middens may also contain flash cubes, babies' binkies, paper, and other trash. Some of the stuff in the midden is the woodrat's trash (nut shells) or food to be eaten later (green plants, acorns), but some has no known use to the rat. Why would a woodrat want a flashcube? Perhaps the midden is an alarm system; a predator or competitor can't sneak up easily if it has to walk over noisy debris.[49] Or maybe woodrats are like some human Mammoth Cave enthusiasts—curious critters that collect cave memorabilia.

Flying Residents: Bats

About one thousand different species of bats in many genera and families make up the order Chiroptera, which means "hand

Gray bats hanging out.

wing." Chiroptera has two suborders, Megachiroptera (megabats) and Microchiroptera (microbats). Megabats tend to be bigger than the microbats. All American bats are microbats.

Prior to white-nose syndrome (more on that topic later), biologists estimated that two thousand to three thousand bats lived in Mammoth Cave. That's not many bats for such a long cave, but in the past the cave was a very large bat hibernaculum. Dr. Merlin Tuttle of Bat Conservation International looked at bat stain—dark stains on the limestone where bats hung, similar to the polish where many people touch rocks—in Little Bat Avenue and Rafinesque Hall in 1997 and estimated that as many as nine to thirteen million bats hibernated there in the past.[50]

Bats also live in other caves, trees, and structures in the park.

Echolocation

Contrary to the old saying "blind as a bat," bats can see. But on dark nights and in caves, they rely on echolocation (sonar) to navigate. In echolocation, a bat uses its mouth or nose to make high-frequency sounds that humans can't hear. If the sound hits something, it echoes back to the bat. The bat can tell the distance, size, shape, texture, and speed of the object based on the echo and thus can avoid it or eat it.

All microbats have echolocation, but, with a couple exceptions, megabats do not.[51]

How Bats Know When It's Night

Most animals know day from night by the sun. Many bats live in trees or buildings in the summer, so they can see the sun go down, but in the cave it looks like night all the time, so how do cave bats know when it's dark outside?

Several things may cue the bats that it's time to get up. The previous night's meal of insects is digested, tummies are empty, so hungry bats wake up.

Perhaps bats wake up when they've had enough sleep. The length of days changes from spring to fall, but bats adjust as nights get shorter or longer.

Colonial bats may rely on social cues. Some bats roost near enough to entrances to see it getting dark. Bats farther back in the cave may hear the entrance-dwelling bats flying or vocalizing, which signals them to get up for dinner.[52] The tricolored bats frequently seen in Mammoth Cave roost solo, so this method probably doesn't work for them.

Bat Diseases

Rabies. Some people see bats as scary and dangerous (although in China they're considered good luck), but bats don't deserve their bad reputation; they're mostly harmless. That being said, they are wild animals, and thus, like other mammals, they can carry rabies.

Dogs cause most of the human deaths from rabies worldwide,[53] mostly in Asia and Africa.[54] Thanks to rabies vaccines for dogs and to pre- and postexposure vaccines for people, human deaths from rabies in the United States have dropped from more than one hundred a year in 1900 to one or two a year nowadays.[55] Bats cause most cases of rabies in humans in the United States.[56] But very few bats have rabies, and they seldom bite people. If a bat is on the ground or behaving oddly, you should assume it is sick and leave it alone—it might be one of those few that do have rabies.

Histoplasmosis. Histoplasmosis ("histo," for short) is a respiratory disease caused by a fungus that grows on bat and bird droppings. People can get histo from breathing airborne spores stirred up from the soil. To get histo, you have to breathe the spores yourself; you can't catch it from someone else who has it.

Many infected people experience no symptoms and don't even know they're infected. If symptoms arise, they include fever, chest pains, and coughing. Mild symptoms usually go away without treatment; extreme cases can be treated with antifungal medication. People with heavy spore exposure can become quite sick and die if not treated; babies, the elderly, and people with lung disease, cancer, or AIDS are most at risk.[57]

Up to 90 percent of people raised in Kentucky and surrounding states have histo before reaching adulthood.[58] People who are exposed to the fungus over a long period tend to build up a resistance to it.

Because Mammoth doesn't have a large bat population, it is no more (and probably less) likely to be a habitat for histo than the surface in Kentucky.[59]

White-nose syndrome. In February 2006, a caver in a New York cave took a photo of a bat with white fungus on its nose. The following winter scientists saw that bats with the white fungus were dying and coined the term *white-nose syndrome* (WNS). WNS kills as many as 90 percent of the bats in caves that it affects. It has spread through much of the eastern United States and Canada.

National Park Service scientists found bats with WNS in remote parts of Mammoth Cave in the winter of 2012–2013. By the following winter, WNS had spread to the cave's tour routes. By the time you are reading this, Mammoth Cave's bat population may be down to two to three hundred individuals instead of two to three thousand.

To fight WNS, scientists are looking at the characteristics of bats that are less susceptible to the disease, such as the Rafinesque big-eared bat (*Corynorhinus rafinesquii*), and trying to find ways to apply those traits to susceptible bats. But even if scientists find a cure or deterrent (Rafinesque big-eared bat microbes, maybe?), it is difficult to treat wild animals: you can't tell bats to come to the clinic for vaccines.[60]

If WNS continues to spread, we will see severe consequences, not just for bats but also for people because bats prey on night-flying insects. As the bat population goes down, the nocturnal flying-insect population goes up.

Surface Animals and Caves

Almost any animal can accidentally fall down a pit and end up confused, injured, hungry, or dead in a cave. Lacking light and plants, caves have little to offer animals, making them unlikely to survive. But sometimes something in the cave tempts them to venture from the sunny, food-laden surface world into a dark cave, and they live to get back out.

Raccoons

Dark stains on the ceiling in Rafinesque Hall indicate that many bats used to hang there, staining the rock with their oily feet in the same way people's hands stain rocks on tour routes. Raccoon scat likewise indicates that raccoons (*Procyon lotor*) visited the same area. Thousands of tasty sleeping bats would make a quarter-of-a-mile trip into the cave worthwhile for a hungry raccoon.

Raccoons are not built for cave life, but scat, claw marks,

Raccoons prefer the surface to the cave. (Courtesy National Park Service)

and other raccoon signs away from entrances indicate they do visit the cave. We don't know how they find their way, maybe by using their sense of smell. Some raccoons that made it into the cave did not make it out; a passage called Fox Avenue was named for a raccoon skeleton found there that was mistaken for a fox skeleton.

Deer

Deer Lick Cave in the park has salt deposits near the entrance. Deer hair caught in the low ceiling reveals that deer visit the cave to lick the salt—hence, the name.

Frogs

The damp Frozen Niagara Entrance and similar entrances provide frogs with crickets to eat and a hideout from most predators.

A damp cave entrance suits this frog as long as it can get food.

Birds

Black vultures (*Coragyps atratus*) often nest at cave entrances but don't go deep into the cave. Dixon Cave near Mammoth Cave's Historic Entrance is popular with vultures. Vultures don't build a nest; they scratch up the dirt and lay their eggs on the ground, so the cave entrance provides both ground and shelter.[61]

Phoebes (*Sayornis phoebe*) tend to build nests on rock ledges, cliffs, cave entrances, and man-made structures that simulate rocky walls.[62] These rocky crags provide shelter and a niche not used by many other birds. Like the vultures, the phoebes occupy just the entrance.

What about animals that we think should like caves but don't? Contrary to popular beliefs, several types of animals do not inhabit caves.

Snakes

Snakes are scary, and caves are scary, so it stands to reason that snakes love caves, at least in some people's minds. Snakes, however, do not live in Mammoth Cave (or in other temperate caves) because it's too cold for them. Lack of food and light also makes caves unappealing to snakes. Snakes can come near entrances or get washed down sinkholes. But when this happens, they become lethargic; 54°F feels like fall to them—time to hibernate. These snakes will eventually die.

In Malaysia, where caves have a warm, snake-friendly temperature, snakes called cave racers (*Elaphe taeniura*) do live deep in caves, where they eat bats and swiftlets.

Bears

Kentucky has a small but growing black bear (*Ursus americanus*) population. Bear sightings near the park have increased in recent years, but no bears have been confirmed in the park. Even when bears used to live here, they did not venture far into the cave. Bears tend to den in small spaces,[63] seldom in large caves,[64] though bear beds (wallowed-out spaces where bears laid down) have been found in nearby Dennison's Cave.

Glow Worms

Caves in New Zealand and Australia are famous for their glow worms. No such critter lives at Mammoth Cave, but if you want to see "glow worms" in a cave, you should head south of the Tennessee state line to Hazard Cave (a rock shelter rather than a true cave) in Pickett State Park: actually, they are glowing fungus gnat larvae, but the name "glow larvae" lacks appeal.

5

The Gypsum-Twinkie Connection

And Other Uses of Cave Minerals

Gypsum

What do the Egyptian pyramids, drywall, hot-dog buns, beer, dentures, and Mammoth Cave have in common? They all contain gypsum.

Gypsum (a.k.a. calcium sulfate) and other minerals found in Mammoth Cave have been used for an assortment of products through the ages. These minerals currently aren't mined out of Mammoth or other caves, though; what's in Mammoth stays in Mammoth.

In Construction

Gypsum has long been used to make plaster. Ancient Egyptians used it for the mortar in the pyramids at Giza.[1]

The moisture in gypsum makes gypsum fire resistant. The fire resistance made gypsum plaster a popular cover for walls of wooden houses in eighteenth-century Paris.[2] The city gave the plaster its name—plaster of Paris. We still use gypsum board, also known as "drywall" or "wallboard," to cover the interior walls in buildings.

Ancient Greeks used gypsum as windows before the invention of glass.[3]

In the Garden

Gardeners use gypsum to condition lawns and gardens. It raises the soil's pH, making it less acidic.

Drinking It

Breweries put gypsum in wort (unfermented beer). It is a nutrient for the yeast, gives the beer a bitter hop taste, adjusts acidity, and prevents haze in the beer.[4]

The Twinkie Connection

Twinkies and other Hostess products, including Ding-Dongs and Ho Hos, contain gypsum (or used to). I contacted the Hostess Company and the Interstate Brand Company, where Hostess products are made; they said gypsum makes the batter light and fluffy.[5]

When the new Hostess Brands took over the bankrupt Hostess in 2013, I checked the ingredients on a box of new Twinkies;

Gypsum: we look at it, eat it, drink it, and put it on our walls. (Courtesy National Park Service)

gypsum is no longer listed, though other Hostess products still have it. I contacted Hostess Brands to find out why it stopped using this seemingly important ingredient, but I have yet to learn the answer.

What Gypsum Is Not Used For

Rumors of toothpaste containing gypsum abound, but after reading all the toothpaste labels in a Winn-Dixie grocery store (yes, really) and finding no brands that contain gypsum or calcium sulfate, I contacted Procter & Gamble, the makers of Crest toothpaste, and Colgate-Palmolive, the makers of Colgate toothpaste. Both companies said none of their toothpastes contain gypsum.[6]

I was still curious about the possibility of gypsum being used in toothpaste in the past. Good fortune brought me in contact with a cave explorer who worked for Procter & Gamble. He contacted several of the company's toothpaste people. They said gypsum is "not very compatible with fluoride" but may have been "used in very early toothpaste powders."[7] In a quest for early tooth-powder ingredients, I found only one mention of gypsum: Persians used burned snail, oyster shells, and gypsum to clean their teeth around 1000 C.E.[8]

One brand of toothpaste does include a cave mineral: some of Tom's of Maine's toothpastes contain calcium carbonate, the main mineral in limestone.[9]

Where Is Gypsum Found?

Some gypsum is synthetically created. Limestone scrubbers on smokestacks at coal-fired power plants capture sulfur dioxide and convert it into calcium sulfate.[10] By keeping the air cleaner, we create a substitute for a nonrenewable resource. Natural gypsum is also mined in Oklahoma, Utah, and Arizona.[11]

Saltpeter

Without saltpeter, people in the late 1700s and early 1800s probably wouldn't have bothered with Mammoth Cave. Saltpeter is the primary ingredient in black gunpowder, invented in China around the ninth century.[12] By the late 1800s, smokeless gunpowder made from nitrocellulose or nitroglycerin or both replaced black powder for regular use.[13] Some hunters and history buffs still use black-powder guns as a hobby (see chapter 8, "Saltpeter: An Explosive Subject," for more on saltpeter).

Epsomite, the Laxative That Softens Your Skin and Repels Raccoons

Prehistoric cavers may have mined epsomite (magnesium sulfate) from Mammoth Cave. Epsomite, or Epsom salts, is an FDA-approved laxative today; whether prehistoric people used it for that purpose or not, we don't know. The Epsom Salt Industry Council claims Epsom salts are good for all sorts of things, including adding body to your hair, soaking your feet, making bath crystals, cleaning tile, dislodging blackheads, treating splinters and scrapes, keeping raccoons away, caring for lawns and gardens, relieving stress, removing hairspray, and softening skin.[14]

Another Laxative

In Mammoth Cave near a rock called "the Devil's Looking Glass," you can see a snowy-looking mineral called mirabilite (sodium sulfate). Archaeologists think prehistoric cavers collected mirabilite by sweeping it off of rocks. Commonly called "Glauber's salts," mirabilite has been used historically as a laxative. Lewis and Clark used Glauber's salts during their expedition.[15]

6

Prairie Park Companion

Mammoth Cave National Park's Prairie Restoration

When you think of Mammoth Cave National Park, you probably first think of caves, maybe forest, but not prairie. But there used to be prairie near Mammoth Cave, and it's been restored. So, you ask, how do we know prairie was there, why did it cease to be prairie, and how did the park restore it?

Evidence of Prairie

How do we know there was prairie near Mammoth Cave? Pollen, archaeological, and historical records indicate there were prairies in the area as well as the forests we see today.

The Pollen Record

In 1981, researchers took a sediment core sample with pollen up to 20,000 years old from Jackson Pond about thirty miles from Mammoth Cave. The sample showed that approximately 3,900 years ago the percentage of grass pollen increased and types of prairie plant pollen appeared, indicating that for thousands of years there were prairie plants in the area as well as the forest vegetation we are familiar with today.[1]

The Archaeological Record

Prehistoric torches found in the cave indicate there was far more open, sunlit ground in the Mammoth Cave area when American Indians used the cave 2,000 to 5,000 years ago. Cane-reed torches in the cave are often one inch in diameter, but you are not likely to find any cane that big growing in the park today; there's not enough sunlight where the cane grows.[2]

False foxglove (*Gerardia flava*) was also common torch material in Mammoth and other local caves.[3] False foxglove needs a great deal of sun to grow and is a parasite on the roots of oak trees,[4] so it's usually found in open oak stands or at the edge of forest openings. Abundant when ancient people were using the cave, false foxglove is rare in the park now. Goldenrod (genus *Solidago*), another common torch plant, also needs a great deal of sunlight to grow, indicating more open meadows when such torches were used.[5]

American Indians made the slippers found in Mammoth Cave from a plant called rattlesnake master (*Eryngium yuccifolium*). A prairie plant now rare in the park, rattlesnake master is common in the meadow at Cave Research Foundation Headquarters at Hamilton Valley just outside the park.

This prairie habitat was most likely created by American Indians lighting fires. Native Americans across the country had many motives for burning areas to create prairie, including but not limited to

- Hunting: Open meadows made good grazing for game animals.
- Crops: Fire created fields for agriculture.
- Improved growth of wild plants: More sun encouraged growth of berries and other edible wild plants.
- Fireproofing: Reducing fuel around settlements protected homes from wildfires.
- War: Open areas had fewer hiding places for the enemy.
- Travel: Open areas were easier to travel through than heavily wooded, brushy land.[6]

The Historical Record

French botanist André Michaux's journal of his scientific exploration of Kentucky from 1793 to 1796 is the earliest known written record of the Barrens of Kentucky, which Barren County was named for. He wrote, "The 12th passed through a Country covered with grass and Oaks which no longer exist as forests, having been burned every year. These lands are called Barren lands although not really sterile."[7]

André's son, François André Michaux, also encountered prairie on an expedition to the Kentucky Barrens in 1802:

> They told me that in this season I should perish with heat and thirst, and that I should not find the least shade the whole of the way. . . . I was agreeably surprised to see a beautiful meadow, where the grass was from two to three feet high. Amidst these pasture lands I discovered a great variety of plants, among which were the *gerardia flava* [false foxglove], or gall of the earth; the *gnaphalium dioicum*, or white plantain; and the *rudbekia purpurea* [coneflower].
>
> . . . The surface of these meadows is generally very even; towards Dripping Spring I observed a lofty eminence, slightly adorned with trees, and bestrewed with enormous rocks, which hang jutting over the main road.[8]

Both false foxglove and coneflower are sun-loving plants now rare in the park. Today the Dripping Spring escarpment is covered with trees rather than "slightly adorned."

Alexander Bullitt mentioned the Barrens in his book *Rambles in the Mammoth Cave* (1845): "The whole route from Elizabethtown to the Cave, passes through what was until recently a Prairie, or, in the language of the country, 'Barrens,' and renders it highly interesting, especially to the botanist from the multitude and variety of flowers with which it abounds during the Spring and Autumn months."[9]

In a plant study published in 1876, John Hussey wrote that buffalo clover (*Trifolium reflexum*) "occurs in several localities between the railroad and Mammoth Cave. . . . I mention it because I have never found so many specimens in any one locality before."[10]

According to Randy Seymour, author of the much more recent book *Wildflowers of Mammoth Cave National Park,* buffalo clover, which requires partial sun, is now "extremely rare" in the park.[11] In *An Excursion to the Mammoth Cave and the Barrens of Kentucky* (1840), Reverend Robert Davidson wrote,

> This tract, extending over several counties, was originally styled the *Barrens,* not from any sterility of soil . . . but because it was a kind of rolling prairie, destitute of timber. While the central parts of the State were covered with forests of heavy timber, or overspread with tall canebrakes, the Barrens with the exception of a few scattered groves along the water-courses, were clothed with a thick growth of prairie grass.
>
> The destitution of timber in the Barrens was owing to the frequent burning of the prairie by hunters to drive out the game.[12]

Reduced Prairie

So why isn't there much prairie in the area now? Reverend Davidson continued, "With the advancing settlement of the country, the prairie fires were gradually extinguished, and the young timber had liberty to grow. The consequence is, that tracts which were destitute of shade ten or twenty years since, are now covered with extensive forests of *Black Jack,* or scrub oak."[13]

Since European American settlement, cleared land has usually been developed into towns, roads, housing, pastures, and farm fields. Undeveloped land is usually allowed to grow into forest rather than maintained as prairie. Until the recent use of prescribed burns, fire was suppressed.

Altering Nature

Isn't the park altering nature by manipulating the land to make prairie? Yes. If people were to disappear altogether, Kentucky would become mostly deciduous forest. But North America has not been a people-free wilderness for tens of thousands of years. Wildlife ecologist Charles E. Kay puts it this way:

> Setting aside an area as "wilderness" or a national park today, and then managing it by "letting-nature-take-its-course," will not preserve some remnant of the past but instead create conditions that have not existed for the last 10,000 years. That is to say, the Americas as first seen by Europeans were not as they had been crafted by God, but as they had been created by native peoples. Unless the importance of aboriginal land management is recognized and modern management practices changed accordingly, our ecosystems will continue to lose the biological diversity and ecological integrity they once had.[14]

The park is trying to achieve pre-Columbian wilderness rather than prehuman wilderness. The plants and animals that settlers wrote about in the 1700s and 1800s had adapted over thousands of years to the ecosystem that Native Americans had influenced.

Cutting Down Trees

Is cutting down trees for prairie the best thing for the environment? To maintain the environment the way it was for thousands of years before European contact and to restore prairie and now rare but once common sun-loving plants, some trees need to be cut down. That doesn't mean a clear-cut, though;

prairies and meadows tend to have some trees. Nor does it mean making prairie everywhere; the land was a mix of woods, savanna, and prairie. Randy Seymour writes about the disappearing prairie plants that

> a very large percentage of the flora in the park is dependent on very small areas of specialized habitats, especially open area type habitats. Despite thousands of historical accounts of barrens and glades as part of the existing landscape pre-European influence and despite the general agreement among botanists that fire dependent barrens were a major component of the landscape pre-European influence, the few remaining barren type habitats in the park were being allowed to disappear along with the often rare flora that were dependent on such areas.[15]

Prairie Restoration and Maintenance

How will the prairie be restored and maintained? North of Flint Ridge Road, the park installed a patch of prairie in the mid-1990s. This disturbed land was graded to create better drainage, and prairie species were planted. Around 2003, at the Great Onyx Job Corps site north of Green River, old farm fields that were growing up in trees and exotic plants were made into a refuge for prairie species. The young trees and exotic plants were removed, and prairie species were planted. The Wondering Woods site north of Highway 255 already had patches of prairie plants in the old fields when it was donated to the park; some trees were cleared to extend the prairie. Wondering Woods is part of Doyle Valley, a huge karst valley that is barely separated from the Sinkhole Plain. This karst valley has many similarities with the Sinkhole Plain and may have been conducive to prairie. In the winter of 2008, Roundstone Native Seed was contracted to clear trees and plant prairie species at Onyx Meadow near Great Onyx Cave. This site

Park ecologist Rick Olson with plume grass on the prairie near Diamond Caverns. (Photograph by the author)

was one of the places in the park where prairie plants have persisted for years.

The most visible prairie restoration is along Highway 255 between Park City and Diamond Caverns. In the fall of 2008,

Robert Poppy Excavating and Construction of Park City was contracted to remove cedar and nonnative loblolly pine trees. The deciduous trees—mostly oaks, black cherry, and some cottonwood—were left. Exotic plants, including Johnson grass, privet hedge, honeysuckle, and multiflora rose, were removed through controlled use of herbicides. The Nature Conservancy has a cooperative agreement with the park to maintain the prairie through mowing and targeted use of herbicide. Prescribed fire will be used to maintain a wooded prairie over the long term.

7

Prehistoric Cavers

At first, the settlers who found Mammoth Cave in the late 1700s thought they discovered a previously unknown cave. They weren't completely wrong; probably no one had been in there for about 1,800 years. However, they soon found torch fragments, gourds, and slippers that revealed that someone had been in the cave long before them. We now know that American Indians left these things behind on their trips into the cave 2,000 to 5,000 years ago. These ancient explorers covered about twelve miles of Mammoth Cave—farther than any other prehistoric people explored any other cave in the world.

Prehistory

We all know dinosaurs were prehistoric, and computers aren't, but just when did prehistory end and history begin? Recorded history refers to time after written language was invented, enabling people to record history. When recorded history began depends what location in the world you're talking about because different cultures developed written language at different times. In most of North America, recorded history began in 1492, with European contact; most native peoples in the Americas did not previously have written language (the Mayan people of Mexico and Central America were the exception).

Cave guide Jackie Wheet and archaeologist George Crothers admire a prehistoric carved walking stick.

Prehistoric People

Were these prehistoric people Native Americans, and what tribe were they? All people in the Americas prior to Columbus's arrival in 1492 were Native Americans. (Scandinavian Leif Erickson and his party landed on North America around 1000 C.E., but their colony didn't last, so they didn't stay or make an impression.) We don't know what the people of the Mammoth Cave area called themselves, but archaeologists call them Late Archaic, Early Woodland people, which refers to the time they lived. The Archaic time period was 6000 to 1000 B.C.E.; the Woodland period was 1000 B.C.E. to 900 C.E.

Cavemen

Were the Native Americans at Mammoth Cave "cavemen"? They lived during the Stone Age, but they did not live deep in-

Ancient cavers used gourds as containers for gypsum and other things they carried to and from the cave. (Courtesy National Park Service)

side Mammoth or other caves, though artifacts indicate they had camps at Salts Entrance and the Historic Entrance.[1]

Even though people from many cultures and time periods have lived in cave entrances, few (if any) lived beyond where sunlight reaches—it's simply too dark.[2]

Human Activity in Mammoth Cave

What were these native peoples doing in Mammoth Cave? Battered gypsum crust and stones worn from beating on other rocks indicate prehistoric cavers were mining gypsum from the walls.[3]

We don't know what they used gypsum for, but making paint and plaster are ancient uses for gypsum, so the Mammoth Cave people may have been doing that.[4]

Along the Violet City route, rocks with no debris other than mirabilite salt make it look as if Native Americans also collected this salt by brushing it off of the rocks.[5] Used sparingly, mirabilite can season or preserve food. If you eat two and two-thirds tablespoons of mirabilite, though, it will act as a laxative![6]

We don't know which use (if either) the ancient cavers put mirabilite salt to.

Archaeologist George Crothers speculates that entering Mammoth Cave to mine or explore may have been a rite of passage for young males. Twelve samples of paleofeces (that is, old poop) have been tested for hormones; they all came from males.[7]

Dating Human Use of the Cave

When were the prehistoric people in Mammoth Cave? A few artifacts have been carbon dated to be about 5,000 years old, but most are between 2,200 and 2,800 years old.[8]

This slipper made of rattlesnake master (a plant not a reptile) is more than two thousand years old. (Courtesy National Park Service)

Native Americans stopped going into the cave about 2,200 years ago, or at least they quit leaving their things in the cave.

How do we know how old the prehistoric artifacts are? Carbon-14 dating (also called C14 dating or radiocarbon dating) is the most common method used to date artifacts. All living things have carbon. When a living thing dies, C14 is no longer produced in it; the remaining carbon just decays; C14 dating measures that decay. It takes 5,730 years (give or take 30 years) for half of the C14 in a dead thing to decay. In another 5,730 years, half of what's left will decay, and so on. This technique is good for measuring objects about 50,000 to 60,000 years old; after that, there's not enough C14 left to measure.[9]

Artifacts on Cave Tours

Visitors on the Discovery Tour see some artifacts on display in cases on Audubon Avenue. All of these artifacts except for the

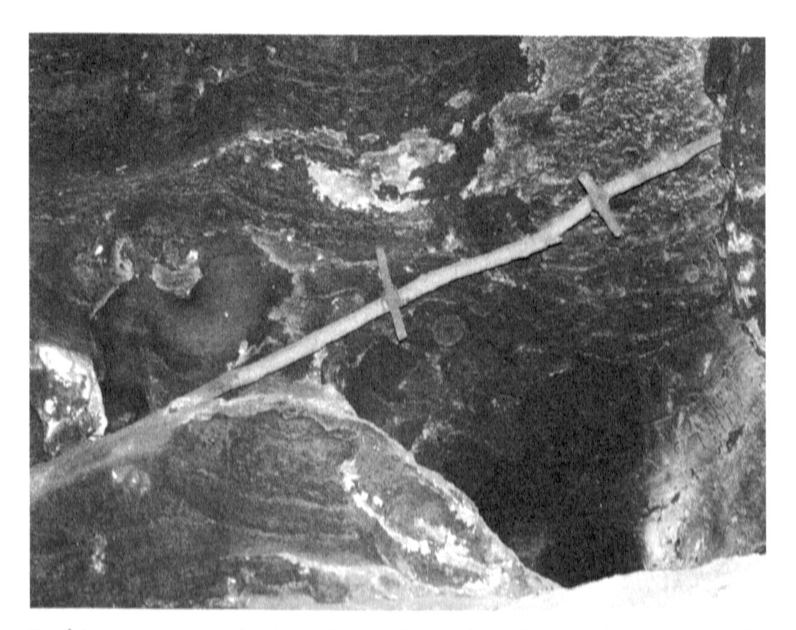

Prehistoric cavers climbed this ladder to high ledges. (Photograph by Bowen, full name unknown)

slippers are authentic; a woman named Joan Miller made reproductions of the slippers for the park in the early 1990s. On Violet City Lantern Tour, visitors can see prehistoric climbing poles and torch fragments.

Several probable prehistoric drawings are also along the Violet City route. I say "probable" because they have not been carbon dated. Scraping the soot off the drawing to date it would ruin it. Even if the soot were dated, that procedure would show only the age of the soot, not the age of the drawing. Although the drawings haven't been dated, they most likely are ancient. Unlike nineteenth-century drawings in the cave, these drawings are abstract—crosshatch patterns, zigzags, and flecks. We don't know what they mean, but they probably symbolized

Ancient drawings cover a rock called Devil's Looking Glass. Unfortunately, the drawings are partly obscured by contributions from more recent visitors. (Courtesy National Park Service)

something to the people who drew them. Several of the drawings are in areas never used for tourism.

Mummification

A few preserved human remains have been found in Mammoth and other caves. Unlike Egyptian mummies, the mummies in the cave were preserved naturally. Even though Mammoth Cave is very humid, there's no liquid water in most parts of it. This lack of moisture is the main reason the bodies became desiccated (dried out) rather than decomposing.[10]

Abandoning the Cave

We don't know why people stopped using the cave. Maybe the gypsum or salt ceased to be valuable enough to make it worth the effort. Time spent in the cave was time not spent hunting or gathering food above ground.

According to archaeologist Guy Prentice, corn was introduced into what is now the eastern United States about the time Mammoth Cave was abandoned. Some ancient Native Americans who grew corn believed caves were entrances to the underworld where spirits lived. Prentice speculates that this idea may have been introduced along with the new food, discouraging people from going into the cave.[11]

8

Saltpeter

An Explosive Subject

Curiosity may have lured the first modern people into Mammoth Cave, but saltpeter, the main ingredient in black gunpowder, kept them coming. Early cave tours began in 1816 partially because cave owners and saltpeter dealers Charles Wilkins and Hyman Gratz wanted to keep making money at the cave after their saltpeter-mining operation ceased to be profitable. More than two hundred years later, the saltpeter-operation artifacts still awe thousands of visitors each year as they listen to guides talk about its role in the War of 1812.

What is saltpeter, though? The word *peter* (or *petre*) means "rock," so the word *saltpeter* means "salt of the rock." Other rock-related words, such as *petrified* and *petroglyph,* have the same root.

For nearly a millennium, people sought saltpeter mostly for gunpowder, but first let's talk about some other saltpeter-related topics.

Saltpeter as Medicine

VISITOR: I hear that military cooks put saltpeter in food to curb soldiers' and sailors' libidos. Does it work?

RANGER: I was in the navy for twenty years, and it hasn't kicked in yet.[1]

We all have heard stories about saltpeter being put in food in the navy (army, prison, boys' boarding school, etc.) to curb men's sexual desires. Caver-physician Dr. Stan Sides told me saltpeter does no such thing.[2] But just because something doesn't work doesn't mean people don't use it. In my quest for facts on saltpeter's use in military kitchens, I contacted Matthew Seelinger of the Army Historical Foundation;[3] Michael J. Crawford, head of the Early History Branch of the Naval Historical Center;[4] and M. G. Kafkalas of the U.S. Army Heritage and Education Center.[5] They all said the military never used saltpeter in food.

It's a good thing soldiers aren't fed saltpeter because if they were to eat too much of it, it would damage their kidneys.

But people in the past did use saltpeter medicinally. The members of the Lewis and Clark Expedition had saltpeter among their medical supplies to treat an assortment of ailments; William Clark wrote, "one man verry sick, Struck with the Sun, Capt. Lewis bled him & gave Niter which has revived him much."[6] (Bonus trivia: William Clark's nephew, Dr. John Croghan, bought Mammoth Cave in the 1830s, and the family owned it until it was purchased by the U.S. government to make it a national park in the 1920s.)

And ancient Romans used saltpeter to treat a buildup of fluid around the testicles.[7] Don't try this at home. Really.

Saltpeter and Gunpowder

The miners at Mammoth Cave mined calcium nitrate. The calcium comes from the limestone, the nitrate from bat guano.

Gunpowder made with calcium nitrate works fine under perfect conditions, but it quickly absorbs moisture, making it too damp to use. Potassium nitrate, or saltpeter, works better. Peter monkeys (a colorful term for saltpeter workers) mixed the "liquor" (the nitrate-rich water leached out of the dirt) with wood ashes, or potash, a good source of potassium hydroxide. As the solution boiled, the calcium and hydroxide joined and settled to the bottom of the pot, leaving the potassium to join with the nitrate—making the potassium nitrate, or saltpeter, as the water boiled off.[8]

The same process (more or less) of converting calcium nitrate into potassium nitrate used at Mammoth Cave had been part of powder making since medieval times.

In the ninth century, the Chinese mixed a concoction of 75 percent saltpeter, 15 percent charcoal, and 10 percent sulfur—the first gunpowder.[9] The little burst of flame and smoke made a good trick, which soon progressed to fireworks. It didn't take the military long to realize gunpowder could be used for more than amusement. Early gunpowder weapons included flaming arrows and crude bombs that catapults flung at the enemy. The Chinese invented the first gun in the late thirteenth century.[10]

In 1267, Englishman Roger Bacon was the first Westerner to write about gunpowder, describing "a child's toy of sound and fire made in various parts of the world with powder of saltpeter, sulfur, and charcoal of hazelwood"—in other words, fireworks. Within eighty years, Europeans had guns.[11]

Bats supplied nitrate-rich guano in caves; livestock and humans did the same job above ground. *Lazarus Ercker's Treatise on Ores and Assaying,* a German book written in 1580, lists old sheep pens, horse stables, rotted garbage dumps, and old latrines as good places to find saltpeter.[12]

The process in which feces and other organic matter decay into nitrate can be artificially hurried along by covering and turning a compost pile—a craft called "petering."[13] Ercker's treatise notes, "The old masters are of the opinion that, if all earth could be brought into sheds under a roof, it would ripen there much sooner."[14]

A spark created mechanically by a matchlock, wheel lock, or flintlock on a gun ignites the sulfur, which ignites the charcoal, which makes the saltpeter release oxygen, which allows the fuel (charcoal and sulfur) to burn more, which makes the gunpowder propel the projectile.[15]

Saltpeter in America

In colonial times in America, most gunpowder was imported from England,[16] though some saltpeter was mined or collected, and a few powder mills existed in the colonies.[17]

Frenchman Eleuthere Irenee (E. I.) du Pont and several members of his extended family landed in Rhode Island on New Year's Day in 1800. E. I.'s father, Pierre Samuel du Pont de Nemours, planned to start a colony called Pontiana in Kentucky, but that didn't work out. E. I. had learned to make powder back in France and felt that starting a gunpowder mill in the United States was more practical than starting a new colony. So in 1802 he built his mill on the Brandywine Creek near Wilmington, Delaware.[18] He dominated the American gunpowder business until smokeless gunpowder replaced black powder in the 1890s.

Saltpeter at Mammoth Cave

In 1798, Valentine Simmons acquired two hundred acres near Green River. A survey of the land done in 1799 shows two unnamed "salt petre caves,"[19] referring to Mammoth Cave's Historic Entrance and Dixon Cave. Someone apparently was already mining or at least saw the potential for mining the cave.

Simmons sold Mammoth Cave (at the time called Big Cave and later Flatt's Cave) to John Flatt, who sold it to brothers George, Leonard, and John McLean around 1808. The McLeans sold it to Charles Wilkins and Fleming Gatewood around 1810 for $3,000.[20] The deeds for all these sales are dated July 9, 1812,[21] which led some twentieth-century history buffs to believe the buyers made some fast, crazy land deals that day. When Flatt and the McLean brothers bought the land, they probably didn't bother with legal papers, but Wilkins and Gatewood wanted no doubt about who owned the land and so eventually made sure the sales were in writing.

Charles Wilkins was an entrepreneur. In addition to being a

saltpeter dealer, he owned a farm, a general store, the Kentucky Mutual Assurance Company, a pig-iron and ironware-casting business; served on a bank's board of directors; and organized the Lexington Fire Department and a state lottery.

Wilkins's brother-in-law Payton Short owned Short Cave, a nearby saltpeter cave where miners found three American Indian mummies. Wilkins displayed one, named "Fawn Hoof," in Mammoth Cave.[22] The publicity from Fawn Hoof may have helped Wilkins turn Mammoth Cave into a tourist attraction when saltpeter ceased to be profitable.

Fleming Gatewood sold his half of the operation to Hyman Gratz for $10,000 in April 1812 or 1813, depending on which source you read.[23] The latter year is more likely because in 1812 miners were making repairs from the New Madrid earthquake that had occurred four months earlier, so mining was slow. The increase in the price of the cave, however, indicates business had picked up by the time Gatewood sold it. Gatewood built another saltpeter operation at Gatewood Saltpetre Cave (now called Hundred Dome Cave) about eight miles south of Mammoth Cave.

Around 1813, Gratz cut Wilkins off from the mining profits even though Wilkins still owned half of the cave. Wilkins died in 1828, and Gratz bought Wilkins's share from his estate for $200.[24] Gatewood and his sons returned to Mammoth Cave to work in the new business of tourism.[25]

The cave's oldest map, the "Eye-Draught" map, drawn before 1811, extends to the intersection of Main Cave with Blue Springs Branch and Blackall Avenue, about two miles from the entrance—way past the mining operation. Perhaps the owners anticipated eventually mining that far into the cave, or maybe they wanted to lead investors to believe the mining operation was more extensive than it was. E. I. du Pont and Dr. Benjamin Rush, a prominent physician and signer of the Declaration of Independence, had copies of the map.[26] It appeared in the 1853 edition of Thomas Jefferson's book *Notes on the State of Virginia.*[27]

The saltpeter vats at Booth's Amphitheater in the cave. (Courtesy National Park Service)

Jefferson had died twenty-seven years earlier, so it's unclear if he knew of Mammoth Cave or had seen the map.

But there is another possible Jefferson–Mammoth connection: in 1806, Jefferson wrote to Pierre du Pont, "The supplies of saltpeter which the Western country can furnish are immense beyond what had been supposed. A single cave is known which would supply us for the whole term of a war."[28]

Was Jefferson referring to Mammoth Cave or another big saltpeter cave in "the Western country"?

A rather cryptic letter to du Pont agent Archibald McCall on January 7, 1813, implies that Mammoth Cave produced better-quality saltpeter than other sources, but at that time the operation was still halted for repairs from the New Madrid earthquake damage.

> *Altho' Mess. Wilkins & Gratz have been Obliged to repair their works,* they have found the means of Selling to the WL last or thereabouts 100,000 lbs of saltpeter; the most

> part of which was coming from their cave we have had an opportunity of ascertaining it by the Saltpetre which we have received Since that time from the Arsnal: and we cannot regret that part of the aforesaid quantity is not by them applied to fulfilling our contract to have the contract completed by Common Kentucky S/petre purchased by Gen al W. would not be very agreeable, because *Some of S/petre is Very bad indeed,* a fact of which you may easily convince yourself by examining the as mixed samples: *we would much prefer to wait Untill Mess W & G have resumed the works of their Cave, which we presume, will take place Spring.*[29]

At the time of the letter, the War of 1812 had been going on for seven months, and it ended two years later in February 1815. During the first months of the war, the mining operation was under repair and unable to supply saltpeter. After workers fixed the equipment, they still had to mine the saltpeter, ship it nearly six hundred miles by horse-drawn vehicle to du Pont in Wilmington, Delaware, manufacture it into gunpowder, and ship it to the army. Although du Pont apparently preferred Mammoth Cave saltpeter, the earthquake damage hurt the operation during the war.

Another cryptic letter from du Pont agent Archibald McCall of Philadelphia in January 1814 implies that Wilkins was no longer in good standing with the company:

> We regret to send by the of y [*sic*] M.E. favour of the 27th Contg. Copy of Chas. Wilkin's letter, That There is no intention from their part to Comply with their Contract . . . That mess W. should have been Careful, when associating with mess G. to make such reserves as wold have enable Them to fulfill contracts entered into by them at a prior period. However Under the existing Circumstances, With a View of avoiding disputes We are of The Same Opinion

> With You that our Bonway is to bring that long Standing Business to a close.[30]

Other Saltpeter Caves and Sources

The cave in "the Western country" Jefferson referred to may have been either Great Saltpeter Cave near Mt. Vernon, Kentucky, or Saltpeter Cave in Carter Caves State Park in Kentucky or Big Bone Cave in Rock Island State Park in Tennessee—all three also had big saltpeter-mining operations in the early 1800s.

Most of the other saltpeter sources were small, but there were many of them. The United States Saltpeter Cave Survey, conducted in 2006 by Douglas Plemons and published in the July–December 2007 issue of the *Journal of Spelean History*, lists 843 saltpeter caves in the United States. After subtracting 78 caves listed as mined during the Civil War, 118 caves that saltpeter historians dispute were really mined for saltpeter, and 21 "lost" caves (caves no one can find, but historic records mention),[31] we are left with 626 saltpeter caves that may have been mined during the War of 1812.

Sandstone rock shelters may have yielded more and better saltpeter than caves. In 1805, Samuel Brown, the owner of Great Saltpeter Cave near Mt. Vernon, Kentucky, stated, "Most of our saltpeter-makers find it in their interest to work the sand rock rather than the calcareous caverns, which yield a mixture of nitrate of pot-ash and nitrate of lime. The rock saltpeter is greatly preferred by our merchants and powder makers and commands a higher price."[32]

Jefferson also recognized the saltpeter potential of rock shelters. In his letter to Pierre du Pont in February 1806, he said, "The caves are numerous. But a more important discovery has been made. That there are immense precipices of a soft sandy rock, when pulverized, yields about 20 pounds of niter, while a bushel of cave soil only produced about one pound."[33]

The War of 1812

So even if Mammoth Cave was just one of many sources of saltpeter (though a big one) during the War of 1812, did the war save the United States from being pulled back into the British Empire?

I asked Dr. Thomas Mackey, a University of Louisville history professor and adviser for the book *War of 1812* by Kelly King Howes. Mackay responded,

> A few of the more extreme politicians in Great Britain talked over their head about regaining the U.S. But Britain's main focus was not the U.S. in the War of 1812, but Napoleon and the French problem. The British considered the American theater a side-show. Yes, they would have taken at least parts of the U.S., but their goal was to break the French-American alliance more than anything.
>
> Only loose talk about retaking the U.S. occurred, not a war goal of the British or a policy.[34]

If Britain didn't want its former colonies back, what was the war about?

To weaken Britain in the war with France, Napoleon issued what he called the Berlin and Milan Decrees, threatening to punish any nations that traded with Britain. Britain retaliated with the Orders in Council, which ordered countries that wanted to trade with France to buy a license from Britain. The United States was neutral in the war between Britain and France, but these decrees made a trade problem for it because it traded with both countries.

British sailors had a tough job fighting France—low pay, harsh discipline, rough living and working conditions—so sailors often deserted and took jobs on American merchant and naval ships. When British navy ships stopped U.S. ships to search for deserters, they sometimes took American sailors along with the deserting British sailors. Stopping U.S. ships and impressing Americans into the British navy increased bad relations between the United States and Britain.

Britain's friendly relations with American Indians defending western land against the encroachment of settlers also caused anti-British feelings.

When President James Madison declared war on June 1, 1812, he stated that the reasons for war were that the British were interfering with trade, impressing U.S. sailors, and aiding American Indians in fighting American settlers.

Faster communication could have stopped the war from beginning. When the United States started threatening war because of interference with trade, British leaders repealed the Orders in Council. Fighting Napoleon was bad enough; they didn't want to fight the Americans, too. But the United States had already declared war before it knew Britain had repealed the Orders.[35]

The End of Saltpeter's Reign

For more than nine hundred years, if you wanted to shoot a gun or blow up something, you used gunpowder made with saltpeter. In 1867, however, Alfred Nobel patented a new explosive made from nitroglycerin; he called his invention "dynamite." It became a favorite with miners and anyone else who wanted to blow up stuff. (Nobel willed most of his fortune to establish the Nobel Peace Prize so that he would be remembered for peace rather than for explosions.) At first, gunpowder manufacturers were against this alternative to their product, but in 1880 the DuPont company began manufacturing dynamite.[36]

Dynamite was tough competition as an explosive, but gunpowder was still used in guns—for a while. In 1886, a Frenchman invented a new, smokeless gunpowder called "Poudre B." A couple years later Nobel improved the new powder by adding nitroglycerin; the British improved it even more by adding petroleum jelly.[37]

By the 1890s, the smokeless, more powerful new gunpowder was more popular than traditional powder. To differentiate it from its modern counterpart, everyone began to refer to old-fashioned gunpowder as "black powder." In 1971, even DuPont ceased making the product that had made the family and the company rich and famous.[38]

Black powder hasn't completely disappeared. Colonial, Revolutionary War, and Civil War reenactors and some hunters use black-powder guns. And we still use black powder more commonly for the same thing its Chinese inventors did—fireworks.

9

The Cave Cure

Old and New Ideas on the Healing Power of Caves

People often associate caves with darkness, mystery, wild animals, and the underworld—scary things. But not everyone sees caves as scary. Many people past and present believe they are places to get healthy.

The Tuberculosis Sanatorium

Tuberculosis is a disease that usually affects the lungs but can attack other body parts. Symptoms include coughing, fatigue, weight loss, loss of appetite, fever, chills, and night sweats. It's not a serious threat in the developed world today but was deadly in the 1800s.

Mammoth Cave owner Dr. John Croghan believed the cave had healing properties, so he set up an underground tuberculosis sanatorium in 1842 hoping to cure the disease, also called TB, consumption, or phthisis.

By January 1843, fifteen to twenty patients lived in the cave, waiting for the cave's healing powers to cure them. In 1843, Dr. Croghan wrote, "I am convinced they would all return to the land above with greatly improved healths."[1]

Things didn't go as the doctor hoped. Some of the patients died in the cave. Those who survived failed to recover, causing Croghan's "resort for invalids," as he called it, to close by the end of that year.

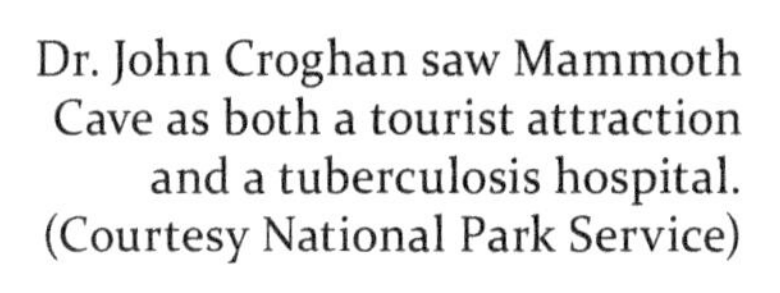

Dr. John Croghan saw Mammoth Cave as both a tourist attraction and a tuberculosis hospital. (Courtesy National Park Service)

Modern cave visitors can still see two of the huts that housed the patients. Most people see the huts as an interesting and unusual piece of history, but some worry about germs. Can modern cave visitors catch tuberculosis more than 170 years after the TB patients left?

The bacterium that causes TB (*Mycobacterium tuberculosis*) tends to survive about eight months.[2] Even if you are around people with TB, it's not highly contagious. You would have to breathe in the germs that an infected person coughed or sneezed into the air.[3]

Healthy people infected with TB might not get sick; their immune systems may protect them. Called "latent TB," this form of the disease isn't contagious, and there are no symptoms. If a person with latent TB gets sick from something else, he may develop active TB because he is more susceptible to it, at which point he will exhibit symptoms and become contagious.[4]

In 1993, Stephen Tankersley, a graduate student at Wright State University, worried some people with his thesis "Detection and Recovery of Anthropogenic Introduced Pathogenic Microorganisms in Mammoth Cave, Mammoth Cave Nation-

Tuberculosis patients lived in huts in the cave. (Courtesy National Speleological Society)

al Park, Kentucky" (English translation: "Finding and Collecting Germs People Brought into Mammoth Cave"). Tankersley stated,

> Samples of soil, water, and wood shavings from a commode were collected in and around the tuberculosis hospital ruins and tested for the presence of the bacterium, *M. tuberculosis*. Using automatable radiometric detection instruments, acid-fast bacterial stains, and conventional culture methods, *M. tuberculosis* was positively identified and determined to be both viable and virulent.
>
> It appears that the climate features of the Mammoth Cave System favor the preservation of perishable microorganisms. . . . As a result . . . *M. tuberculosis* has remained present, viable and possibly virulent in and around the tuberculosis hospital ruins.[5]

In 1995, however, Dr. Joseph H. Bates, M.D., did a follow-up study he titled "Effort to Confirm the Isolation of Mycobacterium Tuberculosis from Soil in Mammoth Cave, Kentucky," but he did not find any TB bacteria. Dr. Bates stated that Tankersley's thesis was "gross incompetence or . . . fraudulent" and that readers "were misled" by Tankersley.[6]

Now that you know you're not going to die from germs left over in the cave from the mid-1800s, you'll be curious to hear more about the cave and TB.

TB Sanatorium Bonus Material!

Some people credit a sanatorium in Gorbersdorf, Germany, founded around 1859 as the world's first TB sanatorium.[7] But the Mammoth Cave sanatorium beat the German one because Dr. Croghan set up his hospital in Mammoth Cave in 1842.

The word *sanare* is Latin for "to heal or cure," so a sanatorium is a place to get cured. There were many TB sanatoriums in the United States and Europe in the late nineteenth and early twentieth centuries. They often exposed patients to nature to provide them with fresh air and nature's other healing properties while isolating the patients from the general population. The sanatorium doctors weren't completely wrong—studies show that being in or even viewing nature is good for our health,[8] but it doesn't cure tuberculosis. By the mid–twentieth century, antibiotics made TB sanatoriums obsolete.

Kentucky has another famous TB sanatorium, Louisville's Waverly Hills Sanatorium, which opened in 1910. TB patients laid in beds on porches, soaked up the sun, and weaved baskets until 1961. After the TB sanatorium closed, the building remained a medical facility that treated other illnesses until 1980.

Today Waverly Hills Sanatorium is a tourist attraction. Some people believe TB patients haunt the building. For $100 (as of 2015), you can spend the night and commune with their ghosts.

Healing Power at Mammoth Cave

Dr. Croghan wasn't the first person to think Mammoth Cave had healing properties. Accounts of great health among slaves who mined saltpeter for gunpowder in Mammoth Cave in the early 1800s may have influenced him.

Ebenezer Meriam visited the Mammoth Cave mining operation during the War of 1812. He wrote, "During the whole time this cave was wrought in for saltpeter, there was no case of sickness among the numerous workmen. They all enjoyed excellent and uninterrupted health."[9]

Robert Montgomery Bird mentioned Mammoth Cave's miners in his book *Peter Pilgrim; or A Rambler's Recollections* in 1838: "The nitre-diggers were a famously healthy set of men: it was a common and humane practice to employ labourers of enfeebled constitutions, who were soon restored to health and strength, though kept at constant labour; and more joyous, merry fellows were never seen. The oxen, of which several were kept, day and night, in the cave hauling the nitrous earth, were after a month or two of toil, in as fine condition for the shambles as if fattened in the stall."[10] In spite of the reports of healthy saltpeter workers, the only miner ever mentioned by name was in fact sick. In 1814, Mammoth Cave manager Archibald Miller wrote to slave owner John Hendrick about a slave leased to work in the cave mining operation, "Your Boy Tambo is very sick and I wish you to come over and see him. . . . I Have bled him Twice and will give him A swett to day I have got no medican at present."[11]

We don't know if Tambo recovered.

Even after the failure of the tuberculosis hospital, people believed the Mammoth Cave air was good for their health. Horace Hovey stated in *Guide Book to the Mammoth Cave* in 1887, "The air is slightly exhilarating and sustains one in a ramble of five or ten hours, so that at its end he is hardly sensible of fatigue."[12]

Dr. Charles Wright wrote in 1860, "Short and easy trips have been known to effect a cure in chronic dysentery and diarrhea, where all other measures had failed. . . . It is not an uncommon occurrence for a person in delicate health to accomplish a journey of twenty miles in the Cave, without suffering from fatigue, who could not be prevailed upon to walk a distance of three miles on the surface of the earth."[13]

In spite of these testimonials, modern park rangers don't recommend cave trips for people with dysentery, diarrhea, or delicate health.

Caves' Healing Powers Worldwide

The cave connection to health goes back a long time. The Chinese wrote about the healing properties of crushed stalactites as early as the fourth century B.C.E. They weren't completely wrong—stalactites are made of calcium carbonate, an ingredient in antacids. They also took powdered stalactites to sedate themselves, to suppress a cough, to stop bleeding, and to encourage milk production in wet nurses.

Europeans also used stalactite powder. In the 1600s (possibly earlier), they took it to strengthen bones and to treat fever. Our bones are made of calcium phosphate, so I see the stalactite-calcium-bone connection, but the fever treatment probably didn't work. They also made a poultice from powdered mondmilch, another calcium carbonate cave formation, to treat people and animals for eye diseases, wounds, and mange as well as to get rid of evil spirits.

In Europe, taking powdered cave formations medicinally faded out in the 1700s, but some people in China still use "dragons' teeth" (fossils from various animals) from caves as heart medicine.[14]

Some people today still believe that caves and cave minerals have the power to heal. They believe that selenite, a type of gypsum, clears the mind, eliminates negative energy, relieves stress and anxiety, and even reduces the risk of cancer. Fans of the healing properties of selenite recommend meditating with it.

In old uranium mines near Boulder, Montana, people seek health instead of uranium. In the Free Enterprise, Merry Widow, and Earth Angel Mines, people with various ailments sit and read, play cards, or chat while surrounded by radon gas, which they believe to be the healing agent in these "health mines." The radon level at the Free Enterprise Mine fluctuates between 1,100 and 2,700 picocuries (the unit of measurement for radon);[15] the level at Mammoth Cave is between 60 and 400 picocuries.[16] As of 2015, a two-hour session at the Free Enterprise Mine cost $16, and a two-hour trip in Mammoth Cave was $15, so your money buys more radon at the mine, but Mammoth Cave is more fun.

The Free Enterprise Radon Health Mine has testimonials from guests (human, canine, and feline), who say the radon helped or cured their arthritis, carpal tunnel syndrome, lupus, multiple sclerosis, and other disorders. A testimonial from Irving the cat states, "I'd like to tell all felines to go to the Mine. It sure changed my life around. My humans are doing great too!"[17]

Radon is listed as a carcinogen by the U.S. Environmental Protection Agency, the World Health Organization, the U.S. National Academy of Sciences, and other health organizations. The National Academy of Sciences BEIR VI Report estimates that 15,000 to 22,000 people die of lung cancer due to radon every year.[18] The concern about radon is bad for business at the radon mines; visitation at Free Enterprise Mine dropped from 5,000 people per summer season before 1978 to 400 people in 2000.[19]

In the United States, radon tends to be viewed as more dangerous than healthy. Sitting in a mine to expose yourself to radon for your health is thought of as (at best) harmless but worthless or (at worst) downright unhealthy. But in Europe some doctors consider exposure to radon (or other things mentioned later) in caves to be legitimate medical treatment; they call such treatment "speleotherapy." The hypothesis is that negative ions from the radon reduce inflammation in the airways, which allows easier breathing for asthma patients.[20] Aggtelek National Park in Hungary offers not only recreational cave tours for tourists but also speleotherapy for children with asthma in Beke Cave, which was declared a medicinal cave in 1965. Speleotherapy is also available in Szemlohegyi Cave in Budapest.[21] Dr. Tibor Horvath of the Department for Speleotherapy and Respiratory Rehabilitation at the Municipal Hospital in Topolca, Hungary, states that speleotherapy patients show long-lasting improvement, reduced request for medicines, and less need for hospitalization.[22]

A study of asthmatic children at Cave Javoricko and Zlaté Hory Mine in the Czech Republic in 1999 showed 60 percent of the patients needed less medication and missed less school after one or more years of speleotherapy.[23]

Gasteiner Heilstollen (Gastein Healing Gallery) in Austria charges 513 Euros (about $570 as of 2015) for three weeks of the "Classical Healing Gallery Cure" in a radon-filled mine turned spa. Austrian or German National Health Insurance covers part of the cost.[24]

Salt caves in eastern Europe are also used to promote health. The salt, like radon, is believed to give off beneficial negative ions. In Armenia, people suffering from allergies, asthma, and respiratory problems ride an elevator seven hundred feet underground into a salt cave called "Republican Speleotherapeutical Hospital" (at seven hundred feet underground, it may be a mine rather than a true cave). The hospital's director and chief doctor, Andranick Voskanyan, says, "The salt environment has an amazing healing impact on the respiratory system." While taking in the healing properties of salt, patients visit with each other, play ping-pong, stroll along the cave passages, and exercise. Treatment lasts three to nine hours a day, five days a week for a month. Before the collapse of the Soviet Union and government health care, the hospital was very popular; now few Armenians can afford it.[25]

The underground Allergologic Hospital in the Ukraine and the Troilus Mine in Romania also provide speleotherapy.[26]

The Versme Health Resort in Lithuania provides an artificial salt "cave" created by covering the walls of a room with salt and blowing salt aerosol into the room for a treatment called "halotherapy."[27] You can buy salt lamps that produce the same effect and make your home into your own private salt cave.[28]

Other factors believed to contribute to the healing power of caves are the stable temperature, high humidity, lack of air pollution, and relatively high amount of carbon dioxide, which some believe encourages deeper breathing and calms spasms.[29]

Evidence shows that cave air can reduce the severity of asthma for many people, but is the benefit worth the negative effect of radon? Do radon, salt, or carbon dioxide in caves and mines cure or help other ailments? Evidence that caves help in the treatment of illnesses other than asthma is lacking. But we know that caves tend to lack pollen and other allergens that plague us above ground. We also know that caving is good exercise, fun, and rewarding—which makes us feel good. Considering these factors, perhaps we can say that caves may have some healing properties after all.

10

Exploring the World's Longest Cave

CAVE VISITOR QUESTION: Do robots explore Mammoth Cave?
OLD ANSWER: No, the technology doesn't exist yet.
NEW ANSWER: No, the technology exists (more later), but cavers can't afford robots.
QUESTION: How about dogs, do they explore the cave?
ANSWER FOR NOW AND FOREVER: No.

Even in our high-tech age of machines that take us to the ocean floor and into space, to explore a cave you still have to strap on the boots and the helmet, get down on your hands and knees, and crawl. No robots, no cave dogs—just you, your gear, and your fellow cavers.

Early Exploration

The first people to explore Mammoth Cave were Late Archaic, Early Woodland Native Americans. Given that they covered about twelve miles of the cave, they were great cavers. But since prehistoric cavers didn't survey, draw maps (as far as we know), or write accounts, we can jump directly to recorded history.

Nineteenth-century guides Stephen Bishop, Mat Bransford, and other slaves discovered areas that are now on tour routes, including Echo River, Mammoth Dome, and Cleaveland

Mammoth Cave guide and explorer Stephen Bishop (1821–1857). Drawing by Bonnie Curnock. (Courtesy Bonnie Curnock)

Avenue, as well as remote passages never used for tours. Stephen Bishop had no survey equipment or cartography training, but he did have impressive caving and mapping skills. In 1845, he drew a map of the cave mostly from memory and from his knowledge of older cave maps he was able to view while staying

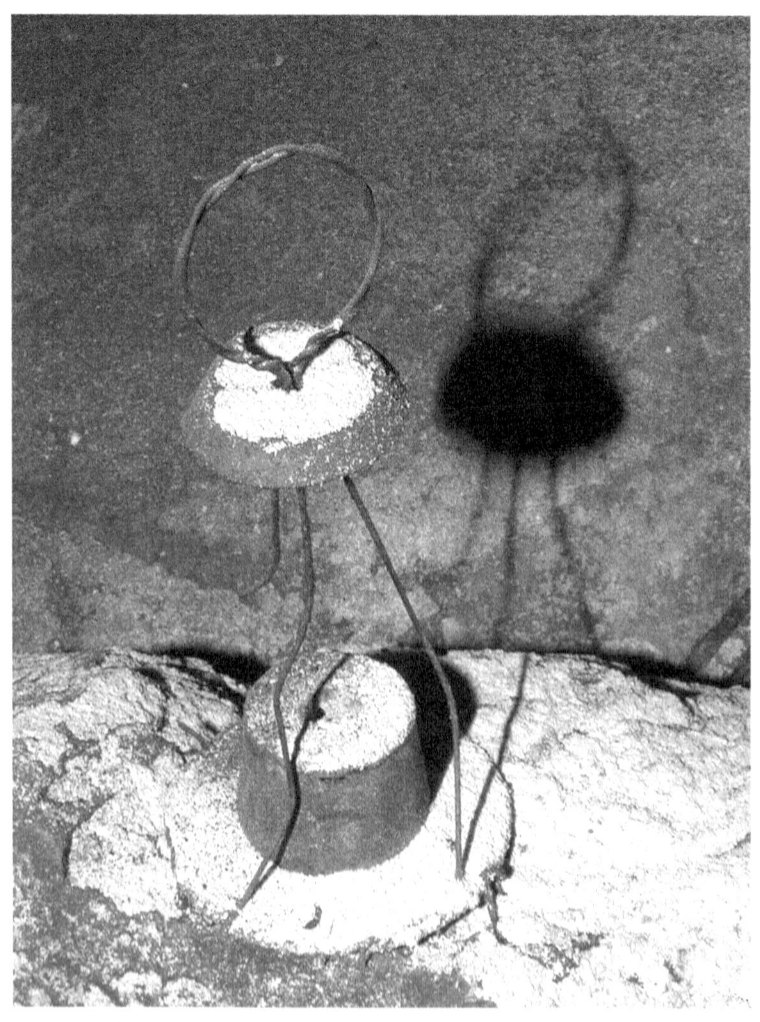

Lighting for early cave explorers. (Courtesy National Park Service)

at Locust Grove, his master John Croghan's home in Louisville. Though Bishop's map is not as accurate as surveyed maps, the passage relationships are correct, and you can find your way through the cave with it.

Many cavers and cave scientists consider Frenchman Edouard Alfred Martel (1859–1938) to be the father of modern speleology—the study of caves. He wrote books on speleology and developed caving equipment and techniques that made previously inaccessible caves accessible. Martel explored America's most famous cave—that's Mammoth—and made profile maps of it in 1912. He commented on the lack of what he regarded as an important caving supply: "I felt most irritated by the strict application of the Anti-alcohol laws in the dry State of Kentucky. Two long days spent in a damp, dark giant catacomb, with nothing to drink other than aqua simplex ('pure water' would be a misnomer), herb tea or ginger beer (which tastes like sugared pepper). . . . A horror to European speleologists! Without a small bottle of rum from my personal luggage I could never have finished this very strenuous visit to Hovey's Cathedrals."[1]

I personally don't recommend rum for caving!

In 1908, German visitor Max Kämper explored and surveyed the cave with guide Ed Bishop to make a map that was more extensive and accurate than existing cave maps. Only portions of Kämper's map were made public (management at the time may have worried that the map would reveal that the cave extended beyond Mammoth Cave Estate property). The map wound up in National Park Service possession but was forgotten until hydrologist Jim Quinlan found it in park files in 1962. The Kämper map was important to the Cave Research Foundation's exploration and survey of Mammoth Cave in the 1960s and 1970s. Caver Diana Daunt copied the original map in 1981, making the reproduction of the Kämper map that cavers and visitors use today.

The Beginning of the Cave Research Foundation

In the early 1950s, Bill Austin, whose family owned Crystal Cave, realized they would soon have to sell the cave to the National Park Service. The land was an in-holding within the park—that is, private property inside the national park. He hoped to increase the Crystal Cave's value with an expedition that showed the cave's extensiveness. (In 1972, cavers discovered a connection between Crystal and Mammoth.) The Collins Crystal Cave or C-3 Expedition brought together a team of cavers that spent a week surveying from an underground base camp in Floyd's Lost Passage in Crystal Cave. The expedition received a great deal of publicity, and a book was even written about it, *The Caves Beyond: The Story of Floyd Collins' Crystal Cave Exploration* (1975) by Joe Lawrence Jr. and Roger Brucker, but the cavers didn't discover much. The real significance of the C-3 Expedition is that it brought together Bill Austin, John Lehrberger, Philip Smith, Roger Brucker, and Roger McClure, who together founded the Cave Research Foundation (CRF) in 1957.

A small shedlike building affectionately called the Spelehut located near Crystal Cave served as CRF's first headquarters. Today, CRF's national headquarters is on Hamilton Valley Road near Mammoth Cave National Park.

CRF promotes research and exploration in caves, cave conservation, and interpretation of caves to the public. Today, it does all the exploration at Mammoth Cave and other caves in the park as well as other caves across the country.

Exploration Today

Volunteers make up CRF. Members don't have to be expert cavers, but they do need some caving background. People with no caving experience who are interested in joining CRF can join a caving organization such as the National Speleological Society to get some experience before moving on to more challenging caving with CRF at Mammoth. The National Park Service doesn't provide any funding for CRF explorations; cavers supply their own caving equipment and transportation. CRF has survey equipment, though many cavers bring their own.

Paleontologist Mona Colburn squeezes through a tight crawl.

There are usually ten expeditions a year at Mammoth Cave National Park, many on holiday weekends.

What's an Expedition Like?

Cavers come to Mammoth Cave from all over the United States, but many are from the Midwest. Those from out of town stay at the CRF headquarters at Hamilton Valley, where they get a bunk and hot meals. Cavers pay for their food and lodging.

On an expedition, several trips go to different locations in the cave. Trips usually have three or four cavers and generally last between eight and twenty hours. These twenty hours are spent working all day and night, not camping in the cave, though cavers may take short catnaps. Carrying just the bare necessities and doing as much work as you can without sleeping tends to be more efficient than hauling sleeping bags, cook stoves, extra food, and other camping gear to set up for a multiday trip. In addition to unexplored areas in Mammoth Cave, trips go to smaller caves in the park and to previously explored areas that need resurveying due to errors made in previous surveys.

People above ground at Hamilton Valley serve as surface watch. If a caving party doesn't return on time, people on surface watch can check the entrance the party used and begin rescue procedures if needed. Since CRF started exploration in 1957, no explorers have needed rescuing.

Surveying

Exploration involves surveying either to map virgin passages or to correct errors on old surveys. Cavers know where new leads are waiting to be explored because when the previous survey team went through that area, they noted side passages that divide off the passage they were surveying.

Cavers often move quickly at the beginning and end of a trip to get through previously explored areas but very slowly in the middle of the trip to survey the new area. If it takes one minute to walk through an area, it can take one hour to survey it, or longer if the passage is complicated to navigate or sketch.

Survey Instruments

In addition to basic cave gear needed for any cave trip (boots, helmet, light, first aid, pack), explorers need survey instruments.

Compass. The compass tells you which direction and what degree you are facing (for example, due north is 0 degrees and due west is 270 degrees). The needle points to the magnetic north. The compass works the same underground as above ground.

Clinometer. A clinometer shows how much of a slope you are on and can be used to measure the heights of trees, buildings, or cliffs.

Measuring tape or laser measurer. The tape or laser measures the distance between survey stations (more on stations later). The laser is easier to use than a tape to measure the distance to the ceiling or down a pit.

But do cavers also make use of high-tech stuff?

Global positioning system (GPS). GPS technology doesn't work in caves because the satellite signal that communicates with a GPS unit can't penetrate the cave.

Cell phones and radios. The cell phone or radio tower signal can't penetrate the rock, either. Very shallow passages can be an exception. One day I overheard radio traffic in the park visitor center. Instead of two voices conversing, I heard only one, saying, "How's that? Can you hear me now?" One caver was in a cave passage called Olive's Bower talking to another caver in the woods. The cavers could hear each other, but I heard only the person on the surface.

Robots or drones. Cave-exploring robots used to be limited to science fiction, but now small drones capable of mapping exist. They would probably work in caves, but their high cost may prohibit common use.

High-tech methods to find voids. High-tech methods are used in cave country to find voids that could influence where to build heavy structures. They are also used to find oil, electric lines, or pipes. These methods don't work well for cave exploration, though, because they can't detect small passages or passages more than one hundred feet deep. Even if you do detect a shallow passage using one of these methods, you only know there is a void, not its details. These methods include microgravity (low gravity indicates low-density material, such as air or water), electroresistivity (rock has high resistivity to current, soil has low resistivity), and ground-penetrating radar (radar waves react differently to rock than to soil).

Survey Positions

Members of a survey team do different jobs.

Lead tape. Your job is to walk (crawl, climb, wiggle) ahead with the measuring tape until the passage bends, meets a side passage, or the tape runs out. Where you stop becomes the next

Another happy caver.

survey station. Each station has a letter and number. If you are calling your survey Q, and the previous station was Q8, you are now at Q9.

Instruments. You read the compass and clinometer to get measurements.

Book. Doing book takes special skill. You not only record the measurements your partners call to you but also sketch a map of the cave as you go.

What Happens after Explorers Leave the Cave?

After returning to CRF headquarters at Hamilton Valley, cavers let the surface watch know the party is back, wash up, and eat a hot meal. Once the trip leaders are well fed, they write a report to describe the trip. The data from the trip will later be entered into a computer, and the sketch made during the survey will be scanned to be used as the basis for a map. The data are jointly managed by Mammoth Cave National Park and CRF. The cave maps belong to CRF.

Thanks to my husband, Rickard A. Olson, a CRF caver since 1973, for his caver knowledge.

11

The Cave Wars

Kentucky's cave country was struck by war in the early to mid-twentieth century: a cave war. In December 1940, the *Louisville Courier-Journal* mentioned the "'battle of the caves' that shook a large section of the Green River country with bitter rivalry and 'sent tourists away cussing Kentucky and everybody in it.'"[1] The war had started many years earlier: an ad in an Evansville, Indiana, newspaper in the early 1920s said, "With solicitors and others stationed on automobile routes, representatives of other caves are threatening to 'put old Mammoth cave out of business.'"[2]

The "solicitors" (here referring to individuals soliciting business for the caves, not to lawyers!), sometimes called "cappers," fought for tourist dollars on the front lines. Who were the cappers, and what did they do? To get their stories, I talked to Joe Duvall, a longtime Mammoth Cave guide who also worked at Diamond Caverns as a guide and capper in the early 1950s; George McCombs, who worked at Great Onyx Cave from 1953 to 1955; and Louie B. Nunn, a Mammoth Onyx Cave guide in the 1940s who became governor of Kentucky from 1967 to 1971.[3]

Cappers worked at information booths along the highway, where they stopped cars to let tourists know which show cave they should see—and which one they shouldn't.

> Sometimes we were called "cappers" because we wore a truck driver's cap with a bill. We tried to get people to go

After his stint at Diamond Caverns, Joe Duval worked at Mammoth Cave National Park from 1961 to 2014. (Courtesy National Park Service)

to Diamond Caverns, naturally. We compared Diamond with the Historic Trip [at Mammoth Cave.] We would tell visitors, "At Mammoth Cave it's history, it's a lantern trip, it isn't well lit, there's a big crowd. If you want a quick cave, there's Diamond Caverns, it's well lit, it's in the cave region, it was formed just like Mammoth Cave, it's more scenic, and it's just a mile down the road. We'd advise you to see that cave and get back on the road quickly." We didn't tell them all the trips that were available at Mammoth Cave.

—Joe Duvall

When a car would come through, she [Lucy Cox, the owner of Great Onyx Cave] wanted us to step out like they had to stop, motion them in, and wait for them to roll down their glass. You had this spiel and a brochure, and if you had the chance, you'd badmouth Mammoth Cave. She said, "Try to keep them away from Mammoth Cave, send them to Great Onyx." When I was working there, Perry [Lucy's husband] and Lucy hated Mammoth Cave. They talked all the time about Dr. Pohl [manager of Mammoth Onyx, Hidden River, and Crystal Caves] and Dr. Rowsey [owner of Diamond Caverns]. There was a lot of animosity. The War of the Caves.

—George McCombs

We praised our three caves, Mammoth Onyx, Hidden River, and Crystal, and usually described only the old entrance to Mammoth Cave and having to wait for scheduled tours.

—Louie B. Nunn

The cappers' stands looked like general information booths for the whole cave area rather than for just a single cave or group of caves under one owner.

Diamond Caverns was not mentioned. We had a drive-in for two cars; the sign over it said, "Official Cave Information."

—Joe Duvall

The best I can remember was a little sign that said, "Cave Information." I'm 90 percent sure there was nothing that said "Great Onyx" because they wanted you to think it was official Mammoth Cave Park information.

—George McCombs

If a car didn't stop for information, a capper could try a different approach.

I had a badge I could wear on my cap or chest. It said, "Official Cave Information." We were dressed in gray pants and a gray shirt, probably to look somewhat like an official ranger.

Sometimes if business was slow and nobody was stopping, we'd step out in the road with a folder and go through the motion of taking down their license number. They'd see us in the mirror and say, "That guy's taking down my license number, maybe I was supposed to stop." They'd turn around and come back. They'd say, "Sorry, I didn't mean to run by you." I'd say, "OK, I'm here to direct you into the park, help you, and give you official information about the cave, the park, and the area." Most of them were apologetic.

—Joe Duvall

If you wouldn't stop, some of the guys would yell at you like a policeman. Some people would stop and back up, thinking they had to stop. I just couldn't do that.

—George McCombs

In spite of the misleading uniform, most cappers were honest.

I don't think cappers were as bad as you hear stories about. But they weren't as good as some people thought they were, either. Sometimes they might stretch the truth a little because they didn't have very many cars that day. We were out to make the owner money; that was my instruction.

We slightly misled people, but we did not tell any wild stories, like [about] diseases and rock falls. We did tell people Diamond Caverns is much more beautiful. And in my opinion it was and still is.

—Joe Duvall

We were never told to misrepresent anything about our caves or others. The Pohls believed they had more to offer

> for the price and permitted us to make our presentation.
>
> —Louie B. Nunn

> I never led people to believe they were at Mammoth Cave. That might have happened, but I don't remember it happening when I was there. A lot of people did think they were at Mammoth Cave; they'd stop on the road, and there was all this propaganda.
>
> —George McCombs

Some people didn't like being stopped; others appreciated the information cappers gave them.

> Many times they thought stopping was compulsory because a guy came out with a badge on. After they found out we worked for Diamond Caverns, a certain percent[age] of them were irate and went straight on to Mammoth Cave.
>
> Solicitation sometimes left a bad taste in some people's mouths, stopping you and trying to get you to go to a particular cave. But it was an old method used in the cave country for years and years and years.
>
> Sometimes after people went through Diamond Caverns, they would stop at the cappers' stand on the way back and say, "We really had a good rip, we enjoyed it, we appreciate your information, we're going to come back next year." A lot of people did come back. Occasionally a car came back and said, 'We went on to Mammoth Cave, and it was great. We didn't want to see your little cave; we're glad we didn't stop." We got both sides.
>
> —Joe Duvall

The cave wars inspired Mammoth Cave to put up a sign that may have caused a problem for park rangers.

For years on the intersection by the stone motel on Highway 70, there was this huge sign that said, "No one has the right to stop you on this road." I don't know who put that up, if it was Mammoth Cave or some privately owned cave, but that sign stayed there forever. There were stories that some older couples would come into Mammoth Cave Park, make some traffic violation; rangers would try to stop them, and they'd speed up and run on through because the sign said you don't have to stop for anybody.

—George McCombs

Joe and George didn't enjoy soliciting, and locals had mixed feelings about cappers.

I never wore a cap. I don't like to wear headgear. I thought it degraded me. I'd be referred to as the lower level, as a solicitor, a capper.

—Joe Duvall

I was working one day, and Lucy Cox said, "George, I want you to come with me." She didn't say what she wanted me to do. Out on Highway 70 they had that little place where they pulled people off the road and solicited them to go to Great Onyx Cave. On the way there she told me what she was going to have me do; I didn't want to do it. I hated the idea of standing out there and flagging people in. She wanted you to be real authoritative about it, act like they had to come in. You had this spiel, and you had the brochure, and if you had the chance, you'd badmouth Mammoth Cave.

I told her, "Miss Cox, I don't want to do this." She said, 'So-and-So's quit, just do it for today." I stayed that day, and I think it was the most miserable day I've ever had in my life. She picked me up early that afternoon, and when I got in the car, I said, "Miss Cox, if this is the only thing you have for me to do, when I get to the hotel, I'm going to pack my clothes and go home. I'm not doing this another day." She said, "That's alright."

Some of them [the other employees] didn't seem to mind, I guess just different personalities. But I would not have gone back out there another day.

—George McCombs

Some people in the community might have been ashamed of the cappers because they did somewhat mislead people. My parents didn't think a thing about it. Those private caves put a lot of money in the area, made jobs for people, and they deposited money at the bank. But a lot of people, the higher social economic level, college educated, looked down on cappers a little. Making your living at the expense of someone else's interest. Sort of like a salesman taking advantage of you.

—Joe Duvall

Cappers and cave guides didn't make much money.

I was not paid by the amount of people I solicited. I got paid a flat rate. The first year I worked there I got paid $20 a week for six days, about $3 a day.

—Joe Duvall

Lucy Cox paid us $4 a day. I used to want to go to the secretary of labor and say, "Are we being paid enough for all the hours we work?" But I never did do it.

—George McCombs

Cave owners made sure cappers did their job, and sometimes they even solicited cars.

As a solicitor, it was awful easy to loaf, not do anything. That's why the owner of Diamond Caverns, Dr. Elwood

Rowsey, asked us to keep track of the cars. If you worked the front of Diamond Caverns, you also recorded the cars that went by—there was a 1949 Plymouth, blue, that went by from California. If the solicitor was just down there and all those cars passed by, the owner knew. "Hey, all those cars went by, and nobody stopped them at the soliciting stand, nobody talked to them. They're not doing anything today." So every night I had this sheet where I recorded the cars that went by Diamond Caverns so he could see if his cappers were doing anything. I thought that was a pretty good idea, to check on us.

—Joe Duvall

Lucy would have men out there soliciting—I know this is the truth—she would take Perry and come from the cave. Perry would bring her up the hill, let her out, and Lucy would climb up a bank, get on a bluff, and watch to see what her men were doing.

A lot of times Lucy would go somewhere, and when she came back, if a tourist was behind her, she would pull up just past her road and stop and tell them to go to Great Onyx. I know that for a fact.

—George McCombs

Though Lucy Cox owned and ran Great Onyx Cave, most cave guides and all cappers were men.

There were no women. I don't know of a single lady that was a capper. But Diamond did have lady guides, and Crystal Cave had lady guides, long before Mammoth did.

—Joe Duvall

Locals called the competition between show caves the "cave wars" for a reason.

We competed for the tourist dollar, but after hours when we saw each other, we were friends.

We all had an agreement; we gave each other free passes. If you were from Mammoth Onyx and you came by Diamond Caverns, we wouldn't make a special trip for you, but you could tag along on any trip. We'd let you and your friends through for free. We'd do that at all the caves.

[But] sometimes, even though nobody was ever indicted or prosecuted, signs directing you to Diamond or Great Onyx Cave would be sawed down at night or burned up. Soliciting stands would have the windows broken, [and] the post holding up the roof would be knocked over. We never did know who did that, but it could have been competitors.

—Joe Duvall

It certainly wasn't friendly competition, I guarantee you.

—George McCombs

Soliciting drew customers to caves even when the cappers didn't talk directly to them.

Sometimes the newspapers would have correspondents come by, listen to the solicitor, and put [what he said] in the paper. We'd make the paper, every solicitor. Sometimes they'd come by and snap your picture.

—Joe Duvall

Mammoth Cave had its own stand.

At that time [the 1950s], Mammoth Cave had an information booth at Sand Cave parking lot. That's why that parking lot is there. Everybody who went to Mammoth Cave pulled in there. They gave them a cave schedule and a folder. Sometimes people who had already been solic-

> ited went by [the booth]. That made the park rangers really mad.
>
> —Joe Duvall

Soliciting decreased in the 1960s. Great Onyx Cave and Floyd Collins's Crystal Cave closed when Mammoth Cave National Park bought them in 1961; Mammoth Cave's information booth soon closed; and Diamond Caverns eventually closed its cappers' stand because other caves abandoned the practice. The Great Onyx cappers' booth on Highway 70 where George McCombs spent "the most miserable day" of his life sat abandoned for nearly forty years and was torn down in 1999.

Cave competition is friendlier today. Most show caves' owners see other attractions as helping attract more visitors to the area rather than as stealing customers away.

12

Want a 2:00 A.M. Cave Tour? Come to Great Onyx!

In 2000, along with his cave wars stories, George McCombs shared with me what it was like working at Great Onyx Cave in the 1950s and the effect that the creation of Mammoth Cave National Park had on his family.

A former in-holding within Mammoth Cave National Park—that is, private property inside the national park—Great Onyx Cave is now part of the park.

GEORGE: I have mixed feelings about the cave. I've always loved caves. I love Mammoth Cave and Great Onyx. But being from Edmonson County—I was born down in Edmonson County—a lot of Edmonson Countians really hate the caves. There are a lot of bad feelings between people in Edmonson County and the caves, especially Mammoth Cave.

When I was a sophomore in high school, I got a job, [or] I was always looking for a job. There were six boys in my family and three girls, and I was always looking for a way to earn a little extra money. I got a job cleaning a restaurant in the morning down in Brownsville. I'd go down in the morning before school, and one morning I'd sweep, the next morning I'd mop, five days a week. When I got to be a junior, I made up my mind, "I'm going to get me a better job," and the first thing I thought of was caves. Down at the

Great Onyx Cave. (Photograph by Arthur Palmer)

high school in Brownsville, we used to have what we called Senior Days. You got a couple days off your senior year to look for a job. Some of us went down to Mammoth Cave to look for a job. At the old hotel on the end next to the visitor center, there was a restaurant and a coffee shop, and on the back side there was a gift shop, and then you went through some double doors into the hotel part. We asked around to see where you could put in an application, and they told us to go through those doors. We wandered around in there and finally found an office with a secretary. It was Garner Hanson's office. [*Garner Hanson was the longtime president of National Park Concessions, the concessioner who ran the hotel at Mammoth Cave and several other national parks.*] I talked to the secretary and told her I was looking for a job. She got three or four of us together and had us fill out an application. I want you to know how green I was: I thought I was applying for a job at Mammoth Cave. I didn't know that was concessions; I had no idea there was a difference. He [Hanson] took our applications and told us he didn't need anybody right now. That was my first defeat with the caves; I didn't get a job. That was my senior year.

My junior year I decided I wanted to try to get a job at some of the privately owned caves—that was before I tried Mammoth Cave. There was a man who lived about three miles from my house who worked at Mammoth Cave. I couldn't get a ride up to his house, so I walked. We didn't have a telephone, and he didn't have a telephone. I walked three miles up to his house. His name was Paul Ray. I said, "Paul, I'd like to ride up to Mammoth Cave with you in the morning; I'm going to try to get a job at some of the privately owned caves." He said, "I'll tell you what: if you're not here at seven o'clock, I'll go without you, and if you get a job over there and want to ride with me, you're going to have to pay me." Well, I didn't like the way he talked. I walked home. The next morning I got up, did all my chores, and walked the three miles to his house. I was there long before seven o'clock, and I rode to Mammoth Cave with him. How far do you guess it is from Mammoth Cave over to Great Onyx? Five miles? I started walking from Mammoth Cave to Great Onyx Cave to put in an application. I was walking along, and an old Ford station wagon came by and picked me up. The fella driving the station wagon was named Jack Gore. I didn't realize it, but he worked at Great Onyx Cave and had worked for Lucy Cox's father, L. P. Edwards. He said, "Where you goin', boy?" I said, "Over to Great Onyx Cave to try to get a job." "Well, I can tell you right now they don't need you." At Great Onyx Cave, I went in to talk to Miss Cox, Lucy, and she hired me just like that. I stayed around that day and made some tours with them and gave myself enough time to walk back to Mammoth Cave to catch Paul Ray [for a ride] back home.

I made up my mind I wasn't going to ride with Paul Ray

anymore; I didn't like the way he talked to me. I told my daddy I got a job at Great Onyx Cave and was going to be staying over there. All the guides stayed at the hotel, except a couple married ones; they got to go home. I went to go talk to Mr. Blanton; he was a ranger at Mammoth Cave. He had a son, Morris Blanton; he retired up there just a few years ago. He got to be a pretty important person up there. I didn't have a driver's license, but I knew how to drive. Daddy said, "Just take the car and go on over to Mr. Blanton and see if you can go to work with him tomorrow." So I started out. He lived pretty close to the park boundary, back by Houchins Ferry. There was a sharp curve, this square turn, [and] I met Mr. Blanton at that turn; he was on the wrong side of the road, and I hit his car. It did about $60 in damage to his car. So I got my job at Great Onyx Cave, and I owed $60 to fix his car; I was already $60 in debt. So the next morning Mr. Blanton picked me up and took me back to Mammoth Cave. I walked again to Great Onyx Cave and started my job.

They told me I could trail for a while. Going through the cave with the other guides, I found out they just had a lot of jokes. They'd show things you could imagine on the ceiling. It seems the tourists didn't care much about that. There was one place where there were two perturbing rocks: you could move a flashlight, turn off all the other lights, move it, and the rocks would come together, and it would look like two people kissing. They had all these corny—I thought they were corny—jokes. You get back by Edward's Dome, and there's a circle in the ceiling. The guide would ask, "Can you tell me what that circle is?" [The guides,] they'd wait and wait and finally say, "That's the ring in the bathtub after your Saturday night bath." Things like that.

COLLEEN: Do you think people wanted to know more about science in the cave?

GEORGE: Yeah, and you didn't have to be a scientist. You could tell them about how the onyx and gypsum were formed.

Lucy Cox, owner and protector of Great Onyx Cave. (Courtesy National Park Service)

I asked Miss Cox if I could change my tour around a little bit; I didn't know anything. She said, "Yeah, that would be alright." She didn't care as long as she got the money. So I asked questions, read a little bit. I made up my own whole spiel, and people seemed to like it. I did away with all the corny jokes. We had a lot of fun over there.

COLLEEN: I hear Miss Lucy was real big on not letting people touch things, break things off, or damage the cave.

GEORGE: Colleen, they should build a statue, a monument, to Lucy Cox. She absolutely single-handedly preserved that cave over there. She threatened us to within an inch of our lives if we let anyone destroy that cave. And we didn't. I turned people around and walked them out of that cave if they were going to handle something. Just turn them around and walk them out, that was it. I'm sorry Mammoth Cave didn't have that foresight.

Since they've been showing Great Onyx Cave, I've gone back over there twice. I thought the guide said some pretty derogatory things about the Coxes and the guides.

We used to have the Complete Trip, an hour and a half trip past the rest area and the Last Rose of Summer, that big gypsum flower; we'd come back to the rest area, turn and go down to the river. Some people would take the trip just to the Last Rose of Summer; they didn't want to go down to the river. So Miss Lucy would sell them the hour trip, and a guide would come in and take the ones that were going out back out. I would take the rest down to the river. Some would only buy the hour trip, and when we got to the benches, they would say, "Is there any way we can go on with the rest of the trip?" In the cave there was a crank phone. We'd call up and say, "We have so many people who want to extend their trip and go on down to the river," so they would know up at the hotel; that would keep somebody from coming in after them. The guide on my tour commented that the guides used to steal that money. I worked there for years, and I never saw anyone steal. He said the guides would have people pay them in the cave and [would] put the money in their pocket. I didn't say anything.

But Lucy, she was after the dollar. Lucy was the boss of the family; she was in charge. In a way, we were treated like slaves. We were on call twenty-four hours a day. If a party came in at 1:00 a.m. and wanted to go through the cave, she'd wake us up, and we'd have to take them through the cave.

More times than not it was somebody who'd been drinking. A lot of times it was military people from Fort Knox, four or five soldiers. I used to hate that, but Miss Cox wanted the money.

We had two trips, an hour trip and an hour and a half. If someone balked on the price, Miss Cox would let them take a thirty-minute trip. It wasn't advertised, but she'd let them go with another group to Edward's Pit or Dome, and another guide would have to go in and pick them up at that point and bring them out—just any way to get the money.

COLLEEN: Was it a friendly competition between caves and the employees? I'd heard a story about employees burning down signs to caves.

GEORGE: They probably did, but I don't know of any specific incident. It certainly wasn't friendly competition, I guarantee you. A lot of times Lucy would go somewhere, and when she came back, if there was a tourist behind her, she would pull up just past her road and stop and tell them to go to Great Onyx. I know that for a fact. When cars came into Great Onyx Cave, this is another thing I didn't like, they didn't just pull up, get out, and come in. Some of us had to meet those cars. We took a bumper sticker; it wasn't the kind that stuck on: it had three metal strips that went around the bumper. We put it on the car and waved for them to roll the window down and gave them the spiel about where to come in and buy their tickets. If the car left, Lucy would be out on the front porch wanting to know why. "What happened, where'd they go?" Not everybody wanted to go through the cave. She was something else.

I remember one time I was bringing a group out of the cave, up the hill, which was a long, steep walk. The children in the group were running ahead, playing as kids will do. They came back to the group and said, "Daddy, Daddy, there's a snake up there." I said, "Stay here and let me look." I walked up the pathway, and there was a rattlesnake, the biggest I'd ever seen. As soon as I get close to it, it coils and starts rattling. Some way, I got it off the path, and we went on past. When we go up to the hotel, I said, "Miss Cox, we were coming up the hill, and there was a rattlesnake right in the middle of the path." She said, "Oh, don't tell them it was a rattlesnake; tell them it was a garden snake." I said, "Miss Cox, it's going to be hard to tell them it's not a rattlesnake when it's coiled and rattling." She didn't want any bad information about Great Onyx Cave at all.

Miss Cox was hard to work for, but I had a lot of respect for her; she really watched after that cave. She paid us $4 a day.

COLLEEN: Was that considered good at the time?

GEORGE: Yeah, I thought it was fine; I was happy to have it. I was between my junior and senior year in high school. We stayed over there and got all our meals. For food we had dry beans, mashed potatoes, corn bread, macaroni, that kind of thing. The cook over there had worked there several years. Her name was Via Ritter. When Lucy and Perry would go away on a trip, as soon as they'd hit the road, she'd go back in the kitchen and fix us a bowl of tuna fish salad or something a little special. We always enjoyed that.

COLLEEN: How was your housing?

GEORGE: Some of the guides stayed in hotel rooms, but most of us stayed in the guide shack down at the entrance to the cave. It had a little front porch on it, two rooms; I think it had two beds in each room. The guides were pretty close; we had a lot of fun together. We used to get together at the entrance to the cave at night when business would slow down, and we'd sing. We had one guide who could sing just like Marty Robbins, another who could sing like Bill Monroe. We really enjoyed that.

I finished working that year, and it was time for school to start. Lucy came to me and said, "George, do you think there is any way you could stay an extra week?" I said, "School's starting; I don't want to miss school." She said, "Why don't you ask the principle and see if he'll let you stay a week?" I did, and he said, "Yes, you can stay out a week, but you'd better catch up in a hurry." I stayed an extra week in the fall and helped them out. I worried about it a lot; I was afraid I was going to get behind in school, but I hadn't really missed anything. Then Lucy said, "How'd you like to work for us on weekends?" I forgot to tell you, I'd get one day a week off. I'd get off in the afternoon, and I'd walk all the way from Great Onyx Cave to Brownsville. If somebody didn't pick me up, I'd walk all the way. I'd get off at four o'clock, and I'd get home about six-thirty. I'd always heard there were lots of wild dogs in the national park; I was afraid I'd run into a pack of wild dogs. Miss Lucy said, "Would you come in and work on weekends?" I said, "I'd love to, but I won't have time to walk home. It gets dark so early in the fall; I don't have a car." She said, "I'll have Perry pick you up at school and take you home on Sunday." They did that my whole senior year.

I graduated in '54 and went back and worked part of '55. I went in the army in '57. In October of '57, that's when the Russians put up the Sputnik, and we were absolutely terrified. We were told they could knock out any place in the United States, and we didn't have any defense. I worried about that a lot. Then in December of '57 my father had a severe heart attack. I got a notice from the Red Cross that he wasn't expected to live, and I got a furlough to come home. There was another plane coming in from Germany, a military air transport that crashed. There was a McCombs on that plane who was killed. The list of fatalities came out, [and] everybody saw that [name]; they knew I was coming home, [and] they thought I'd been killed. I finally contacted my parents, so they knew I was OK, but word hadn't spread around. I went to the hospital to visit my dad. Perry and Lucy had heard I'd been killed, so they came to the hospital; they wanted to see my mother and daddy and talk to them. They opened the door and walked in, and I was there. Lucy saw me, and she fainted, just dead away. They had to carry her out of the room.

Guides liked to play tricks on each other. After you pass through all the onyx and get to the dry part of the cave, there was a big thermometer, stayed right on fifty-four degrees. If we were in the cave by ourselves and we heard a group coming, we would go over to the thermometer, take a cigarette lighter, and warm it up to about ninety. There was a big rock there, and we'd hide behind it. The guide would say, "The temperature stays . . . what in the hell is this?" It finally got so everybody was used to it, but if we had a new guide, we'd heat up the thermometer for him.

COLLEEN: What did people think of the river?

GEORGE: There was a story that it wasn't really a river, that it was just a dammed-up puddle, but I don't know. People seemed to like it; I think they liked the idea of being down that far. And the way we went through the cave, it was like going down a ladder; it was almost straight down. There was just a small platform there. It [the river] had the eyeless crayfish and the eyeless fish. We had to take down a few [people] at a time; the platform was so small, only ten or twelve could stand there.

Lucy would forget about the guides if it got busy. She was so interested in getting that money that she'd just forget about us. Lunchtime would come, and we'd be going in and out. You'd come out of the cave, and there'd be a group there waiting for you. One day it had gotten so bad it was two o'clock, and none of us had had lunch. We said, "We're going to get something to eat, or we're not going to make another trip." So we got on the old phone and called up to the hotel: "Miss Cox, we're not going to make another trip until we get some food." People were coming over the hill, and people were waiting to get in the cave, and it was the fastest thing I've ever seen. I don't believe it was ten minutes until the cook and some helpers were coming over the hill with trays of food for us. We got our lunch.

COLLEEN: Were all the guides male?

GEORGE: Yeah, no women guided at all.

COLLEEN: Did Lucy or Perry ever guide?

GEORGE: Never saw them at the entrance to the cave. The only place I ever saw them was at the hotel. Perry was real easy going, didn't worry about much, spent a lot of time sitting on the front porch rocking. Lucy was running here and there trying to get everything organized.

COLLEEN: Did people ever think they were at Mammoth Cave?

GEORGE: Absolutely.

COLLEEN: Did guides let them think they were there?

GEORGE: I never did. That might have happened, but I don't remember it happening when I was there. A lot of people did think they were at Mammoth Cave because they'd stop on the road, and there was all this propaganda.

Lucy's father's name was L. P. Edwards; he discovered the cave [on] June 12, 1915. That's what we said at the entrance. But the story goes he didn't really discover the cave, that it was discovered on his farm. There was a fellow by the name of Edmond Turner who came into the Mammoth Cave area. He was a civil engineer, a real small man. Someone said he was so small he looked sickly, but he wasn't. He got interested in the caves and wanted to do some exploring. So he asked around to find somebody he could get with and do some exploring. The name that kept coming up was Floyd Collins. So he got together with Floyd Collins, and they did some exploring together. I can't remember if they were exploring Colossal Cave or Salts Cave; I believe it was Salts. They said somebody around there had a still where they made peach brandy. It was a pretty good distance from L. P. Edwards's farm, but there was a spring on his farm, and these peach pits kept showing up in this spring. That gave them the idea that there must be a cave where the water was carrying these peach pits through. The story goes that Turner approached Edwards with the information that there was a cave on the farm and worked out a deal with him to try and find the cave; he was supposed to be a business partner. They did find the cave; Edmond Turner's the one who found it, but for some reason he got pushed out and didn't get anything out of it. The story goes—and I've heard this from several different people—Edmond Turner already knew Great

Onyx Cave was there because he'd already been in it from Salts or Colossal Cave. They said he was [so] angry with L. P. Edwards because L. P. pushed him out of the deal that he went into the cave and blocked up all the passages that connected Great Onyx to Colossal. I don't know whether or not that's true.

COLLEEN: I don't suppose Miss Lucy wanted you to tell that story.

GEORGE: Oh no! That was talked about in private.

L. P. Edwards had two daughters named Lucy and Cova. They [the Edwardses] named the river in the cave after them, the Lucy Cova River. Cova married a fellow named Harry Bush. Cova never did have anything to do with the cave that I know of.

COLLEEN: Did Miss Lucy tell you what to say about the cave's discovery?

GEORGE: We were told to say, "This cave was discovered June 12, 1915, by L. P. Edwards. He discovered the cave by walking around the hillside about fifty yards from where we enter. He noticed a stone with some onyx and felt some cool air coming out of a crevice. This gave him an idea there might be a cavern; he sunk a shaft down about twenty feet before he broke into the cavern." We were at a certain point when we told them this. We shined our flashlights back and said, "The old entrance to the cave is back in this direction, and tourists for a long time had to change their clothes and crawl on their hands and knees to get into the cave." Then we explained that they blasted another shaft to make the entrance where we just entered. We were told what to say about that; [about] most of the other stuff we could say what we wanted.

I used to get mad at Miss Lucy because she would call us at five o'clock in the morning. She'd ring that telephone down at the entrance, and we'd go up and eat breakfast and clean up and shave. Then she'd start our chores; she didn't want us to rest. She'd have us whitewashing—all the trees around the hotel were whitewashed so high—she'd have us sweeping the paths, painting, all kinds of dirty work. You'd be all hot, dirty, and sweaty, and some party would come in, [so] you'd have to stop what you were doing and take them through the cave. I really hated that. I thought if you were going to be meeting tourists, you should be staying clean.

Lucy and Perry moved to Cave City. I don't remember which one died first; there was only a month or two in between.

COLLEEN: What did they do after the [National Park Service] bought the cave?

GEORGE: They had some farmland that they oversaw. They were too old to do any physical farm labor at that time. I don't remember what the cave brought, $398,000?

COLLEEN: How many hours a day did you work?

GEORGE: I made as many as seven trips a day. A mile and a half in, mile and a half out, three miles round trip—that's twenty-one miles a day I walked. I'd walk twenty-one miles, and on my day off I'd walk to Brownsville. The thing was, we were on call twenty-four hours a day. The only things furnished were our food, our room, the batteries for our flashlight; we furnished our own flashlights and uniforms. I always tried to wear some sort of uniform, matching khakis or something. A lot of the guys would just wear blue jeans and a flannel shirt [and would] look more like farmhands than guides.

One day a party came over the hill, and they were dissatisfied with everything. One of the guys said, "How in the world do people make a living in this God-forsaken country?" [One of us responded,] "It's not too hard; we just sit around and wait for people like you to come along and pay $6 to see a hole in the ground." That guy didn't say another word.

COLLEEN: Was that the price, $6?

GEORGE: I can't remember exactly how much it was. There were two advertised trips. It was close.

Perry cured his own hams and sold them at the restaurant in the hotel. Sometimes they'd fix us a ham breakfast. That was a big day. Most of the time we'd just have scrambled eggs, biscuits, bacon.

It's kind of a bittersweet thing. I enjoyed the job, enjoyed the guides. Thinking back on all the hard work we did, we really weren't compensated that much. I feel that way about caves. I'm crazy about Mammoth Cave, but another part of me still dislikes it.

Where [Joppa Ridge Motor Trail] comes out on Highway 70, you turn left to get back on 70, just a little turn. The old road didn't turn; it went straight into 70. In that fork is where I was born. My father owned that property. Right in the fork we had a grocery store. I was born in '35; we didn't have to move out of the park until '39. We moved from the park to Brownsville. I was born at Joppa. There was an old high school across the road. My dad taught school up there. He told me stories about when they first got ready to buy the park, that officials—I don't know if it was somebody local who was hired or appointed or officials from Washington—would come to the schools and talk to all the little kids. They'd tell them what wonderful things they were going to do with Mammoth Cave. They were going to build this beautiful place; there was going to be recreation for everybody; "there's going to be jobs for all your fathers, your brothers, and your uncles"; they'd make good money. Then they'd take up a collection from the little kids to help pay for the park. My dad hated that so much. Then when the park was formed, people from Edmonson County got pushed aside. In fact, when I told someone I was going to go to Mammoth Cave to try and get a job, they said, "You're from Edmonson County; you might as well stay home." It was that way for a while; I think it's kind of died down now.

The CCC [Civilian Conservation Corps] camps, one of them was kind of close to us at Joppa. Daddy would sell all kinds of things to the CCC boys. He got a car and started a taxi service. He'd run the CCC boys to Bowling Green, to Glasgow. He was on the road all the time. It got so bad, two of my older brothers took over the driving.

My brothers would tell stories about going down the road, and a park ranger would crowd them off the road and wouldn't stop to help them get back on the road. I heard things like that all the time when I was growing up. That's why I have a hatred—not really a hatred—a distrust for Mammoth Cave.

13

Cave Guide Coy Hanson

Electric Lights, the Mammoth Cave Railroad, and Other Stuff

With Rick Olson and Chuck Decroix

The Hanson family comes up often in twentieth-century history of Mammoth Cave. Carl Hanson electrified the New Entrance; he and his son Pete helped discover the crystal-filled passages called New Discovery; and his son Garner managed National Park Concessions, which ran Mammoth Cave Hotel and concessions in several other national parks for years. Pete Hanson came close to connecting the Mammoth and Flint Ridge cave systems in the 1930s, though he didn't know it. The cave explorers who made the connection in 1972 saw Pete's name carved on a rock in an area of Flint Ridge Cave they thought was unexplored. Seeing that Pete got to Flint Ridge Cave from Mammoth Cave clued the cavers in 1972 that the two caves had to be one—this connection made the system the longest in the world.

Carl's son Coy guided tours at Mammoth Cave from 1938 to the early 1960s and eventually worked at other national parks. Park ecologist Rick Olson and I visited Coy in his home to record his stories on May 18, 2009. Rick and park ranger Chuck Decroix went to see Coy again in 2012.

RICK: Joe Duval said back in the day the New Entrance had a

Delco battery system. [*Blasted in 1921, New Entrance connects to a series of natural pits that visitors descend on three hundred stairs.*]

COY: It did.

RICK: And you worked . . .

COY: Not worked, I was with my dad; I was about four or five years old. But I do remember going to Bowling Green to haul distilled water for those batteries. They had a little building in front of the entrance. There was an array of batteries that lighted the dome section.

RICK: Do you remember what year that was?

COY: 1925. To the best of my knowledge New Entrance opened in 1921—the same year I was born, as a matter of fact.

RICK: Did the New Entrance have the Delco electric lights from the time it was first shown?

COY: I doubt it because I was just four or five years old, and I believe it [the light system] was just in the beginning stages.

RICK: So they showed it by lantern light?

COY: They'd almost have to. You know they had to build those steps down through there with lanterns. There'd be some daredevil people who'd done that. Those big timbers [that made the stairs] had to be sawed with a hand saw.

I lived at Crystal Cave [a section of Mammoth Cave] for a couple of years. After I went from being a guide to being a ranger, my first station was Crystal Cave.

COLLEEN: Did you live in the Collins House?

COY: That little building that was there. My duty once a month was to go down and make sure that Floyd was still OK. [*Cavers and National Park Service people are on a first-name basis with Floyd Collins, who was trapped and killed in nearby Sand Cave in 1925. His body was on display in Crystal Cave until the park purchased the cave from private owners and closed it in 1960. He remained in the cave until the 1980s, when he was buried at the Mammoth Cave Baptist Cemetery.*]

COLLEEN: What do you remember about the old Mammoth Cave Railroad?

COY: It run adjacent to our farm. Our house was just a field away. Then again, I was just a very young man. But I remember it seemed to be two times a day it would go in and out, and it would come again. I know it would jump the track very easily, and people would get wrecking bars and so forth and get it back on.

COLLEEN: Did you see it jump the track?

COY: No, I never did. But some of the outlaws, you might say, would put soap on the track. [*Laughs.*]

COLLEEN: Was it the local kids who put soap on the tracks?

COY: I didn't do anything like that. By the Doyle Valley Cemetery there is an iron gate. There used to be a store there. Roy France's father ran that store. They had a little siding there; they kept a little handcar there—for emergencies, I guess. But, anyway, at night some of us would get on that thing and take a ride. [*Laughs.*] You'd have to work getting it back, but that happened.

COLLEEN: How did you make the handcar go? Did you pump it with your hands?

COY: You have to get off and push mostly to get it going. It rolled pretty easy going downhill, but coming back uphill . . . . But I can remember that train going by very distinctly.

COLLEEN: Did you ever ride the train?

COY: No, I never. And of course it was replaced by the school bus. I did ride on that.

COLLEEN: You rode on the bus when it was on the rails?

COY: Yes. That ran for, it seems like to me, for two or three years after the train quit running. It ran a year or two. I should remember better than that. I'm just eighty-seven years old.

RICK: What was your dad's name?

COY: Carl Hanson.

RICK: He was in on the discovery of New Discovery.

COY: He and my brother and Claude Hunt and Leo Hunt. Dad

had forty-four years with both caves—I mean the same cave with different management. He started in '21, the same year the New Entrance Cave was open. I'm not sure about it, but I think he was instrumental in helping to build those steps.

COLLEEN: Your dad worked for George Morrison? [*Morrison owned the New Entrance section of Mammoth Cave before it became a national park.*]

COY: Yes. George Morrison had a 12-gauge shotgun. My dad borrowed it, and we kept it for many years. Dad used to go hunting with it. But shortly before Mr. Morrison committed suicide, he wrote and asked for his shotgun. He got his shotgun back, but I think he used something else to put an end to his life. I guess in the middle of the big depression, he probably got in pretty bad circumstances. I don't ever remember seeing him, but a lot of people thought he was a pretty good man.

There used to be a building right near New Entrance. They didn't cut down the trees; they just built around them. There was a family; a Mr. Yarrell was the man who operated the cave. He had a brother-in-law, Mr. Braden. But they operated that cave for the first five or six years, I know. Mr. Braden was his brother-in-law. He gave me a little vase shaped like an acorn; it had been run through a lathe. The bottom still had the bark on it, but the top was plain wood. My sister has that still over at Chaumont in her house. I was just a little boy, and he gave me that.

RICK: Was that something he made himself?

COY: He didn't make it around here; it was good lathe work on it. It had a little rim down there, and the top fit down over that rim just perfectly. I have some fond memories of that place.

Mr. Yarrell was a German gentleman, kind of portly. I don't think he went up and down those steps much! [*Laughs.*]

My first public job was being a bellhop at the old New Entrance Hotel. After that building was no longer there, everything operated out of the hotel up on top.

COLLEEN: When you worked at the New Entrance Hotel, was the Park Service in charge?

COY: No, this was in '37 when I started. I don't know if you knew Arthur Doyle or not. He worked at the old Mammoth Cave Hotel as a clerk. He and his wife were managing the New Entrance Hotel at that time.

Coy told Rick and Chuck more stories when they visited in 2012.

COY: In 1937, I was a night bellhop. I was very green. One lady said, "Have you ever thought about getting shoes that don't squeak?" [*Laughs.*] [The year] 1938 is when I started at Mammoth Cave.

CHUCK: As a guide?

COY: As a trailer. I didn't get to be a guide until later.

CHUCK: What were your duties as a trailer?

COY: Keep the lanterns clean, bring them back from Mt. McKinley, and put oil in them. But you didn't put in too much because you had to carry them back to Mt. McKinley! [*Modern trailers bring up the rear on cave tours, handle first-aid needs, and turn out the lights. Thanks to electric lights, they no longer carry lanterns on most tours.*]

CHUCK: That's a long haul.

COY: You're telling me! [*Laughs.*] We'd take gunny sacks—grass sacks, I called them. Take nine lanterns and bail them around one sack. We put the gunny sacks through the lantern handles, put a nail in the sack to keep it together, tie two sacks together, and carry them out.

CHUCK: What was it like doing tours on Echo River?

COY: That was one of my favorite places, Echo River. One episode I recall, on the Fourth of July it rained tremendously. My brother Pete, Marvin Sells, and I were sent down to the river to check out the situation. The river was almost up to the ceiling; there was about a foot of clearance. Pete and

Rick Olson and Coy Hanson show off a photo of Coy Hanson on Echo River. (Courtesy National Park Service)

I put our gas light onto the boat, got in, and felt our way along the ceiling until we got to the fourth landing that had a higher ceiling. We got back in time to tell Mr. Chalet [the boss] to let the All Day Trip go. [*High water like Coy describes could stop boats from getting through low areas. Boat trips on Echo River stopped in 1992 because of the flooding and concerns about possible impact to aquatic cave life.*]

Rick: Oh my goodness! You told me about a little episode in the Corkscrew that you'd never told anyone before. Could you tell that story?

Coy: In those days, the guide on the Echo River trip was responsible for going over to Minnehaha Island to pick up the boat from the All Day Trip and bring it back to the fourth landing. I got back from getting the boat and started up the Corkscrew. There were a lot of steps. When you get to the top of the steps, there's a little chicken walk. It had cleats, but I missed the cleats and fell forward down the

All that's left of the Hanson brothers' homemade basketball hoop.

steps and knocked the mantle off my lantern, but it still put out enough light so I could see. I collected my thoughts, climbed through Corkscrew and came out, turned down the lantern, and went into the Guide House. Nobody even knew it until you did! [*Laughs.*]

RICK: One day I was walking near Union City [*a pre-national park community that's now in the woods*], and I found an old home place that had two chimneys at either end. When you described your place, the light came on; I was at your home place.

COY: If you go back again, look at the cedar trees: you'll see a metal hoop that was our basketball hoop. The tree had fallen; the hoop was still on it.

RICK: I'll look for it. That's wild.

Rick went back to the Hanson home place and found Coy's homemade basketball hoop.

14

Lost John

The Discovery of a Mummy

A fair amount has been written about the archaeology of the naturally mummified Native American "Lost John" discovered in 1935, but the story of how cave guides Lyman Cutliff and Grover Campbell found the mummy and the unusual experience they had while guarding it was oral history for years. I decided to get as close to the source as possible to find out what happened. On May 25, 1997, I visited Lyman's son, Lewis Cutliff, at his home in Park City, Kentucky, and asked him to tell me his dad's story.

COLLEEN: Tell me about your dad's experience in Blue Springs Branch [*a passage near where they found Lost John*].

LEWIS: At that time, they had the CCCs [Civilian Conservation Corps workers] in there working on the trail. He and Grover Campbell were assigned from the guide force to go down and catalog anything they might find. They'd been finding some artifacts like moccasins, burnt sticks, cane reeds, and what have you. They would keep field notes on where that stuff was found, and they had a lot of time on their hands. They did a lot of exploring; [my dad,] he'd been exploring caves since he was little. He'd go into caves with his daddy. They would see things in Salts Cave. They would see things in Mammoth Cave, the same types of things. He knew that his uncle Bill Cutliff found a mummy in Salts Cave. He and Grover started looking around; they decided if there was one in Salts Cave, there might be one in Mammoth Cave.

Daddy was off one day. The next night he came back. Grover had been there the night before by himself with the crew. He told Daddy to come up on this ledge; he wanted to show him something that he had found the day before—some mummified bats up on the ledge. They were up there crawling around. They crawled around this boulder, and Grover put his hand down. It felt funny, and he cursed and said, "What is that?" He moved back, and Daddy came and looked at it and said, "Well, Grover, that's what we've been looking for; that's a mummy's head." They went and reported it to Mr. Charlet [the cave manager]. They immediately started to notify Washington. Alonzo Pond was the archaeologist they sent down here. They started working on that, getting the rock off the mummy.

After they found it, they had them stay there and guard it at night. While they were there, they would look around. They were in Blue Springs Branch digging around. Grover put his hand down in this hole, and they heard three knocks. It scared Grover, and he said, "What was that?" Daddy said, "I'll put my hand in there," so he put his hand in. He said it felt like pumice, real fine powder. He heard three knocks, then three more knocks—like somebody with a rock banging on another rock. It frightened both of them; it was loud enough that they heard it! So they went back up and stayed with the mummy.

It seemed like anything that happened to either one of them was always in threes—for three knocks, then three more. Grover died three years later. Six years after that, Daddy lost his job. Thirty-three years later my sister had open-heart surgery, and she died—no, it was thirty years; then she died in three years.

He always felt like he shouldn't have bothered the mummy. It just felt like that was what caused all the problems we had. He often said if he had it to do over again, he'd never tell it; he wouldn't disturb it.

COLLEEN: Was he still alive when they put the mummy away in the '70s?

LEWIS: Yes, he lived until 1986.

COLLEEN: Was he glad when the mummy was put away?

LEWIS: I feel so; he was always sorry that he disturbed it.

15

CCC Boy Fred K. Hanie Jr.

During the Great Depression of the 1930s, thousands of young men worked for the U.S. Civilian Conservation Corps (CCC) in national parks and other federal land. Known as the "CCC boys," they planted and cut trees and built trails, hiking shelters, and cabins, many of which park visitors still enjoy today.

By the twenty-first century, few CCC boys were still living, but I had the good fortune of meeting Reverend Fred K. Hanie Jr., age eighty-two, on September 15, 2003, at Hopewell Baptist Church in Glasgow, Kentucky, where he pastored for more than forty years. He shared with me stories of being a Mammoth Cave CCC boy.

COLLEEN: What were you and your family doing when you got involved with the CCC at age sixteen?

REV. HANIE: We moved to Glasgow in February 1937. My dad was a chef cook at the old Spotswood Hotel. We, my brothers and sisters, went to school there: it was a two-story frame building.

The summer of '38, some of the fellas who had been at the CCC camps told me about it. I liked it, and I wanted to go. So I went up to a lady—I think her name was Mrs. Briant. She was in an office on the south side of the square. I went up there and told her I wanted to go. She took my name and address. She told me I could go and asked how old I was. I told her I became sixteen [on] April 29. She said, "If your parents

Although CCC teams were usually segregated, this team was integrated. Reverend Hanie worked with the black CCC camp at Mammoth Cave. (Courtesy National Park Service)

will let you go, it will be all right, but if they won't let you go, then you can't go. I'll take your name down, and I'll notify you. You'll be going in to Mammoth Cave." She sent me a card and told me to come up there. She said, "Did you talk to your parents?" I said, "No, but my daddy works at the hotel; I'll go up there now and talk to him." I'd be leaving in about two weeks, so I went up to talk to my dad. He said he wanted me to at least finish high school. He said, "Son, I didn't get to finish high school because we worked all day." I said, "Dad, since you and mom are not together, things are different now. I feel like I need to get out and try to make it for myself. I could learn a lot over there, and that's where I'm going because they give you a dollar a day, and they have barracks, and they give you clothing; your lodging is free and three meals a day. If you let me go over there, I'll be one less mouth that you have to feed." He didn't want to, but he saw I was determined to go. I said, "Furthermore it's just twenty-eight miles from here; I'm close to home. They're not going to send me far away." Some of the fellas went to other places. He let me go. I went over there in September of '38, and I came out September of 1940.

In the time that I was over there, I worked in the woods. During that time, chestnut trees had a blight, so they had crews that would go around across the river and search the woods and find chestnut trees that had that blight. The fellas would mark them, and they'd cut them down. If they [the trees with blight] weren't too far apart, they'd put them together. I was in the crew to come get them. Sometimes they'd be in those bottom places, way downhill. We had those log hooks. Two to four fellas, one on each side, would get those hooks and hook them in a tree and get them [the logs] up the hill and lay them at a certain point all together, and the truck would come by and pick them up. We'd also cut all the limbs off so we could handle them. We did that for quite a while in the summer, the first summer I was over there.

That winter I worked in the rock quarry on fair days. Did you ever see that rock quarry?

COLLEEN: I don't think so. Was the rock quarry to get limestone for gravel?

REV. HANIE: Yes, to make gravel. We had a crusher there. They were blasting on the road that they took to the rock quarry. It was maybe one-fourth of a block from the road to where they blasted. They blasted and blasted and blasted. It was a big wall of limestone. They started from up above at the top. They'd blast all the rock down. After they got all the rock down, they'd go back up and start at another place.

We'd be down in that quarry breaking that rock. We were bare-backed. We had sleeveless undershirts, and we'd take those off. I remember I used a fourteen-pound sledgehammer, and I enjoyed it because they told me how to break those rocks by the grain, and I learned that real quickly. I'd break those rocks and pile them up. In a day's time, I'd pile them up all around me at least four or five feet high. I'd have to leave an exit so I could get out. The truck would come, and the fellas would load them [the broken rocks] up and carry them up to the crusher. We kept that crusher going and kept making more. The trucks would come and haul the gravel away. They used that gravel for different things, but most of that gravel went on the roads. They cut the roads in different places where they needed them for one reason or the other. I worked at the rock quarry one whole winter. We did that in fair weather, if it wasn't too cold.

That second year they asked me if I wanted to do what they called the ECW [Energy Conservation Work] down below the quarters. It was about two blocks from the camp, the way we had our quarters and mess hall. They had quarters down there for the foremen who worked these fellas. My foreman at the rock quarry was Mr. Clark. He used to have a dairy on 31W as you're going to Park City. He had a farm there, and some of his boys may still be over there. He was a real nice man, treated us nice. There was Mr. Clark, and there was Mr. Gibson. I used to know all those foremen's names because they had quarters there. One of them had to be on duty all the time. There were five or six of them. One was scheduled to stay there all night in case an emergency came up, like a fire or something where officials needed to be there. One of them had a room there at their quarters. I used to go there every day and clean up the lobby where they set in the winter and summer months. I'd sweep up and see that there was a fire in the fireplace. There was Mr. Black, Vernon Black; he was superintendent of the working organization. He had five or six bird dogs; he loved to bird hunt. I used to bring food from the kitchen, garbage, about half of a five-gallon bucket, to feed the dogs twice a day, at noon and at night. They had a garage where they brought the trucks to repair them and a blacksmith shop where they did all kinds of repair work to different equipment.

I met some real good people there, real good friends. They were friends after I grew up and the war was over. I even pastored a gentleman who was a brother to a deacon in this church and the Park City Baptist Church. His name was Arthur Bransford. I pastored that church about ten months in Park City. I was his pastor, but I knew him from when I was a boy at Mammoth Cave. He worked down there in the blacksmith's shop.

I remember the last summer we were over there. We had just finished lunch that Sunday when somebody rang the fire bell and told us to fall out in the company street. There were a lot of fellas gone that weekend, but there were a lot of fellas still there, about fifty of us. A fire had started; I don't know the directions now. They got us ready and put us on trucks. We didn't leave until evening, about an hour or so before dark, and we found our way to the place we were supposed to go. We had axes and rakes and cutting utensils. The wind wasn't blowing too hard, but we were supposed to cut out foliage for distance to keep the fire from spreading. We did that until it got dark. On the other side, there were two white camps. [*Black men and white men did the same work and received the same pay in the CCC, but their camps were usually segregated.*]

They were working on that fire because it was pretty widespread. By the time we got there, they said they had discovered the fire that morning, and they'd been fighting it just about all day. When we got there midevening, we started working on the area, keeping it from spreading. We cut a long path in a certain place about the length of this room so it wouldn't spread to the other area because the wind was blowing, so we tried to put a distance between it. After it was all over, they said we'd did a good job, all the fellas who worked over there from all the camps. They said [that] where we'd worked, the fire never got out of proportion that we couldn't fight it.

There was an experience with one of the fellas with us. We were going from one place to the other. Our leaders told us to be careful and stay behind each other and make one path. One boy was doing more talking than he should, and instead of watching what he was doing and listening to orders, he stopped and was hollering at some of the boys way back. It was pitch dark that night; the fellas he was following [had] gone off and left him. They had told us very emphatic to stay behind the person who's in front of you and not to lose sight of that person. He got lost, started trying to catch up with them, and went into a big hole about eighteen feet deep. We were going back and heard somebody say, "Help! Help! Help!" Our leader stopped and said, "Where is that coming from? It's coming from back there." Our leader came back, and he had a flashlight. He said, "You all stay here, and those of us in the rear end of the line (he took four or five of us) will go back and see if we can find him." He [the fellow in the hole] was about as far from us as the front door of the church [is from here]. He was still hollering. Oh, he was afraid, he was about to panic. There were some trees growing down in that hole. It was about as big as my office. They had limbs on them that broke his fall. He was scratched and bruised up pretty bad. There were some small bushes down there; he hit those and that broke his fall. He was a scared fella. When he saw those lights, he said, "Come and get me, you all!" There was a tree there; he couldn't see it, but our leader saw it. He guided him over and said, "Here's a tree." He climbed up it where we could pull him out of that hole without having any extra equipment. We didn't have a rope at that time, though we could have gotten one. This tree was just big enough to hold his weight. He lived in another barracks than what I did, but they said he had nightmares; it took him a long time to forget it. He was a hometown boy from here. When we got out of CCC camp, we used to tease him.

There was a gentleman from Louisville; he was our educational supervisor. Some of the boys here were not too well versed on education. He taught them to write and do arithmetic. He was a musician; we had a glee club. We had about twenty-five boys who could really sing. They'd sing hymns and gospels. On Sundays, trucks took them to the local churches in the country to sing. They helped a lot of people.

We could go to town every Thursday. Sometimes we would go to the movies in Cave City; trucks would carry us down there. We'd go and look at those shoot 'em ups. Some of those fellas I can remember. There was an actor named Bob Steele, and [there was] Tim McCoy. They said Tim McCoy was from over here around Tompkinsville; that was his original home. And there was Tom Mix. Those are all I remember right now, but those I remember real well.

I'd come home as much as I could. Some fellas would come and carry us home and take us back over there for fifty cents apiece. Gas was pretty cheap back then, and they did us a favor. Three or four cars would come over there because they would write to them in advance and tell them how many were coming home. They'd charge fifty cents one way. To carry you back was another fifty cents, so they were making some pretty good money; they'd make three or four trips over there.

COLLEEN: Were there any other activities besides the Glee Club?

REV. HANIE: Yes. They carried us to the cave, and we went on the longest route through Fat Man's Misery and down to Echo River. They had a boat there; it seems like they had two boats. You had to bend over to keep your head from hitting the ceiling. I never shall forget that. Those fellas from the Glee Club sang some gospel songs and hymns, and their voices really sounded good. The ceiling was low, and it seemed to me there was an echo. I'm not certain about that, but it seemed their voices carried beyond where they were singing.

COLLEEN: Yes, there is an echo down there.

REV. HANIE: Their voices were so crystal clear. Their voices were so rich, and they had such harmony.

COLLEEN: Did you ever work in the cave, or was all of your work [for the CCC] in the woods?

REV. HANIE: It was in the woods. But they had other groups, and they worked in the fields to prevent soil erosion; they made field damns where the water was washing the soil away. That group did that every day, and there's no telling how many acres within the two years I was there that they built those field dams.

I remember one thing—this is not too appetizing to me. One of the fellas, he was from Louisville, he killed a five-foot rattlesnake over there in those fields. What happened was they were standing there talking, and this rattlesnake began to rattle. One fella detected it and said, "You all hush your talking; there's a rattlesnake over there some place." It wasn't close enough to strike, but he knew if he didn't call their attention to it, somebody might not be paying attention and walk over close enough to it and get bit. This one boy, he was real tall and about eighteen—we used to call him handsome, he was real handsome, had curly hair—he went over there and said, "Let's see if I can find this rattlesnake." He got a tool, a hoe or something, and went over there and found it in the weeds and killed it. He carried it home with him that weekend. He said he was going to have a belt made out of it, and he did. He had that done in about six months. He brought that belt back and showed it to us.

COLLEEN: I brought some photos that I copied for you. Maybe there will be something you recognize. Here are some fellas out in the woods. It looks like they are clearing some trees, and here they're cutting some wood.

REV. HANIE: This might have been a time they were cutting those dead chestnut trees. See those hooks?

COLLEEN: They're doing the kind of work you did with the chestnut trees?

REV. HANIE: Yes. This might have been the crew that went before us. They'd go and find trees and mark where they were. They'd cut them down and cut the limbs off of them. Then we'd come down. Sometimes we went into very steep places.

We were worried about snakes. We saw some snakes one time. A boy in our group stepped on a bunch of eggs, and somebody took a stick and opened [one] up; there was a little snake in it. I don't know what kind it was, but he come crawling out. It looked like ten or fifteen eggs. When he crawled out, one of the fellas said, "Don't open more; I'm going to burn them up." Someone said, "Just let them stay there, maybe they'll die." He said, "Maybe they'll hatch, too; we won't come back here anymore."

[*Looking at a photo of men standing on the ferry.*] This is a familiar sight, these fellas on the boat.

COLLEEN: Did you often ride the ferry?

REV. HANIE: Yes, ma'am. When we worked across the river, we went down that road by the cave. At that time, there was a ferry down there [by River Styx Spring]. We rode the ferry across the river when we were packing those trees up out of those hollows and places back there.

COLLEEN: Did they have a truck to pick you up on the other side?

REV. HANIE: The truck would ride the ferry across; we stayed on the truck. We were so far away, they'd bring our lunch over there too. We stayed over there all day. In the evening, we'd come back to the camp. This is so clear, such a familiar sight.

COLLEEN: Here's another photo of the fellas cutting trees.

REV. HANIE: Yes. See, there's some wood stacked up.

COLLEEN: Was the idea in cutting the chestnut trees to stop the blight from spreading to other trees?

REV. HANIE: No, I don't think so. They treated those trees to make telephone posts out of them, those that were big enough. That's what they told us. They soaked them in some kind of oil to resist water. Back in those days, you had to do with what you could, so they might have treated them with used motor oil. In later years, I know on the farm they'd use motor oil to treat fence posts. They'd soak them several times to get it in there, put those fence posts in the ground, and they would last for several years.

Another thing, them having that wood stacked up so high [*referring to a photo of CCC boys by a woodpile*]; there's a saw there, too. That's what you call a crosscut saw. We had one on my granddaddy's farm. They may have used that wood to make fires when they were out in the wintertime, fires to warm by. We had woodstoves at the rock quarry. We had a building we could come in and get warm. These are really good [*referring to the photos*].

COLLEEN: Here's a photo of one of the camps. I think that's the camp on Flint Ridge Road. Is that the camp you were in?

REV. HANIE: Flint Ridge? No, the number of our camp was Company 510. This doesn't look like ours. To get to Company 510 you didn't come in contact with anything else; you just go in and turn down that road and go down the hill go for a mile or so. After you got down there, there was a gymnasium. Is that still there?

COLLEEN: No, but I know where the building used to be.

REV. HANIE: They used to go down there and play basketball. You'd go down a piece, turn left, go about a quarter of a mile, and come into Company 510. There were four barracks besides the kitchen and the dining room where we ate. There were some other buildings there. There was a first-aid building, and an educational building; students went to school down there.

COLLEEN: You mentioned the gymnasium. Did you ever go over and play basketball?

REV. HANIE: I didn't, but we had some fellas who did. We had some teams, and they played each other. But I didn't play; I just sat and watched.

16

Ronald Reagan's Visit to Mammoth Cave

Ronald Reagan's visit to Mammoth Cave on July 12, 1984, made the news, but the newspaper articles leave out the best details. To find out what really happened, I interviewed cave guides Keven Neff and George Bruce Corrie at Mammoth Cave National Park on June 25, 2011, to get their stories on Reagan's visit to the cave.

COLLEEN: Tell me about the day Ronald Reagan came to the cave.

KEVEN: I was off that day, but I figured if I put on my uniform, I could walk around, and no one would say anything. So a couple of us walked out to the staging area near where they would bring the choppers down. We were standing there, and one of the marine guards came up and said, "You guys better get into the press box because Reagan's chopper will be coming any time." "Oh, thank you," and we walked over to the press box. Henry Holeman [a law enforcement ranger], bless his heart, took one look at us and said, "What are you guys doing here? You have no clearance to be here, now get." So, tails between our legs, we slinked back to the visitor center.

One of the Secret Service guys came over and said to me, "Would you check out that camper that's parked over there?

Walk by and listen, see if you hear anything inside." Apparently, they thought some snipers might shoot outside the top of the camper. This was Reagan's first public appearance after getting out of the hospital after he'd been shot, so security was pretty tight everywhere.

I'm standing there, and all of a sudden this chopper flies overhead; it swoops over the crowd, [and] hats are going everywhere. There was this great big sign on the wall that said "Edmonson County Republicans Welcome President Reagan"; that sign got torn to shreds with the downwash from the blades. Finally, the Reagan chopper landed, and he got out. He stood on a platform by the press box. He waved to the group, and everybody waved back, and everybody cheered. He went down the steps and got into a black limousine. There were three limousines parked out there. The way I understand it, they crisscrossed their way over to Frozen Niagara, so you really didn't know which limo he was in.

The whole park shut down. They blocked all the roads; you couldn't go in or out.

COLLEEN: Were there any visitors in the park?

KEVEN: Oh yeah. Unfortunately, there was a group coming back from the Great Onyx tour stuck on the side road; they [the Secret Service] wouldn't let them get out of the bus, and it was really hot. So they had a horrible time waiting for all that.

Just as Reagan left, there was a kid standing by his bicycle, leaning on a tree. Just then, the tire blew, a very loud bang! I think half the people in the crowd were Secret Service people because they all start running at me with their hands in their vest pockets like they were going to pull out a gun. I'm standing there in front of everyone, holding up my hands, saying, "It was only a bicycle tire, it was only a bicycle tire!" I wonder if that tire had blown while Reagan was standing there on the platform, they probably would have thrown him right back in the chopper and took off. They wouldn't have known it was strictly a bicycle tire.

I got to see Reagan, and he may have seen me standing there. When he got to the Frozen Niagara Entrance, there was a display of pictures for him to see. Bob Cetera was the photographer at that time, and I was a model. I had a chance to look at the display later; there were about sixteen pictures, and I was in ten of them, so Reagan got to see me, that's for sure.

The way I understand it, he went into the cave, and there were pipes and wires all over because he couldn't be out of sight of the little black box and little red telephone in case there was an attack. He had all his communication. Not only did they have wires running around, [but] they [also] had wireless; they had receivers—it was a mess. They put a big light bar up across the top of the drapery room. Reagan only went down about ten steps from the bottom of the Drapery Room. All the reporters were in the Drapery Room, including Bob Cetera. They were taking pictures of him [Reagan] while he was standing there looking around. I don't see how he could see anything with the bright lights shining in his face and flash bulbs going off.

After he left, he went to Beech Bend; I think they were having a camping convention. He went there to give a speech, too.

Somebody said that when he was in the cave, none of the communication worked. So if the Soviets had wanted to launch a missile at that time, they had about ten minutes. They couldn't get to Reagan because the phones didn't work. When they checked them, they worked, but when they went to tear them down, they checked them again, and they said nothing worked. It's kind of spooky.

COLLEEN: Tell me about checking the KEA Room [a cave chamber] before he arrived.

KEVEN: The KEA Room is right under Frozen Niagara. There are some good formations down there. The sand all along the top of the ledges around the room is as beautiful as all get out—just pristine. One day we went down there after Reagan had visited; somebody had crawled back in and dug all this up, made a total mess of it.

COLLEEN: Was it the Secret Service?

KEVEN: I found out later they had gone down there, and someone said, "Where does that tunnel go? Better check it out," because it comes up under Frozen Niagara. So somebody else did it, but I reported it as destruction, so I guess my name came in that I had crawled back there and checked, but I didn't. A few guides like to spread that rumor that I was elected to crawl back in there. It would have been fun—nothing more interesting than crawling through pristine soil to find out what's there. I don't know what's back there. It seemed kind of strange to me—if it's pristine, nobody would be in there.

It was interesting how much security there was for Reagan. They were here months before [the visit], checking out hotels, motels, campgrounds, walking the road to Frozen Niagara, having dead limbs cut off trees, station[ing] dogs by each culvert in case someone wants to blow up a culvert. It was good security.

Ranger George Bruce Corrie also was at the cave when Reagan came.

GEORGE: A security team came here about a week before the president did; they had to run a background check on all of us. They brought in a Dodge motor home; Dodge at the time made the biggest motor home. It sat at the camp store where they could tie into the phone lines. It provided onsite security and communications for their detail here.

The day before the president came, they sent men into the cave. I think one of them was at the Crystal Lake passage; the other one was at College Heights to secure the area and make sure no one came in the night before to cause security problems for the president. They spent the whole night in the cave.

COLLEEN: Was it just a security person, or was there a Park Service person?

GEORGE: I believe one was Park Service.

COLLEEN: One of the law enforcement rangers?

GEORGE: Yes. When the president went in the cave, a man carried a black box that had a satellite phone.

COLLEEN: Keven told me it worked during the initial testing but did not work when they removed it.

GEORGE: Yes. There were six busloads of people, and two were selected to go to the entrance and be the audience for the president. We gave him a jacket and a hat.

COLLEEN: Did you see him?

GEORGE: Yes. I saw him going in the cave and coming back out. And when he talked to us, he shook some hands, but I didn't get to shake hands with him.

COLLEEN: You were standing at the Frozen Niagara Entrance?

GEORGE: Yes, right by the presidential ivy.

COLLEEN: Was your role to be the audience?

GEORGE: Yes, we were onlookers.

COLLEEN: What did he say when he came out of the cave?

GEORGE: He was in the area to speak to a group in Bowling Green. He talked some about projects in the park and how it was good federal spending. I don't remember the details of what he talked about.

COLLEEN: Did you see the helicopter land, or were you already at the entrance?

GEORGE: I saw it come in, but I was already at the entrance, [so] I didn't see it land. I do remember they had roads blocked off so his caravan would be the only vehicles on the road. I believe a different helicopter went over the caravan to provide coverage. It was amazing how many people participated in what appeared to be a just a moment event on TV. It took days and weeks of planning for this casual visit to the park.

COLLEEN: How did they do the security check on everybody?

GEORGE: I don't know. They checked our employment records.

But I don't know if the FBI checked them or how they did that. It was quite an undertaking.

COLLEEN: Anything else you remember?

GEORGE: No, except we got to name a tour for him.

COLLEEN: The Presidential Tour, now called Frozen Niagara.

17

Rachel Wilson, a Woman in the Underground World of Men

As a female cave guide in an era when women make up about half the Mammoth Cave guide force, I was curious about the experiences of early female park rangers and cave guides in entering what was then considered a man's domain. So I visited Rachel Wilson, the first female permanent National Park Service guide at Mammoth Cave, at her home in Edmonson County, Kentucky, on March 18, 2009. She was almost eighty-one years old when she shared these stories.

COLLEEN: When did you start working at Mammoth Cave?

RACHEL: I started in the spring of '57. I took a civil service test at the park. About twenty of us took the test for a teller position, selling tickets. I got the teller position selling tickets in the old white-frame hotel. Franklin Jolly was a teller, and so was Miss Felishia Velanova; the three of us sold tickets. After that, I was seasonal for about three years. They opened up a permanent position for an information receptionist. Their excuse for hiring me over ten-point veterans—there were veterans available—was that they needed a woman on the force to take care of women visitors, so I got the job as full-time information receptionist. I was required at that time to wear three-inch heels.

COLLEEN: The three-inch heels were official uniform shoes?

RACHEL: Yes, they sure were.

COLLEEN: Hopefully you didn't have to wear them when you started working in the cave.

RACHEL: No, I just worked eight hours behind the desk. When I started working in '57, we didn't have a uniform available to buy. They had the picture of it and knew what it was supposed to be made out of. I ordered material from a tailor in Louisville and had a local seamstress make me a skirt and a jacket. There was no way to order the uniform; they just had the picture.

COLLEEN: Everyone had to have a uniform tailor made?

RACHEL: No, I was the only one, the only woman.

COLLEEN: So the men could order a uniform?

RACHEL: Yes, they had been in the uniform for years. My uniform was a Park Service green, six-gored skirt, white blouse with a shawl collar. That was the first uniform we had; mine was homemade.

COLLEEN: Did you have a hat when you were working information?

RACHEL: I had a hat; it came on later, and it was like a little airline hostess hat. I had to wear it all the time. One day I was over at the hotel at a luncheon. The superintendent's wife was there, and I was there with my little airline hostess hat on. Somebody said something about the hat, and I said, "I hate this little hat; I wish I could take it off." The next day the superintendent came over and said, "You don't have to wear that hat; go ahead and take it off." My supervisor didn't like that; he got real upset, but there wasn't anything he could do! The superintendent's wife had told him I didn't enjoy wearing that hat all the time.

COLLEEN: Do you still have the hat or other pieces of the uniform?

RACHEL: No, when I retired in '80, I gave all my uniforms to Sharon Ganci [another early female guide]. They were a little short on her, I'm only five foot two, and she's about five foot seven! They were short, but she wore them anyway!

COLLEEN: Where did you work before Mammoth Cave?

RACHEL: I worked in Louisville as a bookkeeper for a carpet company on Market Street. Before that I worked in Evansville as a bookkeeper for a wholesale drug company for a number of years.

COLLEEN: What made you want to work at Mammoth Cave?

RACHEL: Unemployment insurance required that I go look for work!

COLLEEN: Mammoth Cave sounded like a good place to work?

RACHEL: I had to fill out places I had looked for work. They called me from the unemployment office and told me there was an opening at Mammoth Cave, that they were going to give a test and to go take it. I went up, took it, and got the job. And I enjoyed it, really enjoyed it. I was happy to get a job, really. Living out in the country, there wasn't any work around. Bowling Green didn't have work back then.

COLLEEN: What made you decide to become a guide?

RACHEL: After the information desk, I went back to selling tickets. From ticket sells, I applied for a clerk's position in the superintendent's office, and I got that. I worked as a clerk two or three years. I never did enjoy it; I always liked the other side, the activities and all.

One day the superintendent called me into his office and said, "We've got an opening for a chief teller, and we want to know if you'll take it." I said, "No, I downgraded and came over here so I could be off Saturdays and Sundays." He said, "If I promise you Saturdays and Sundays off, will you take it?" So I went back as chief teller.

Then there was an opening for a guide. This was when women's lib was just beginning to take over, in the '60s. I went with my application when the assistant chief naturalist came out his door and said [to me], "Are you going down to apply for that guide job?" I said yes. He said, "You might as well turn around and go back; you're not getting it. We're not having a woman on the guide force." So I went back and forgot about it, and they filled the position with some man. It wasn't too much longer until another position came open. The superintendent came over and said, "I understand that you were going to apply before and you were told not to." I said, "That's right." He said, "You put your application in this time." So I did, and I got the job.

COLLEEN: What made you want to apply for the guide's job?

RACHEL: I thought it would be interesting and I would enjoy being out from behind a desk. I knew all the men who were there, and I thought it would be more fun to get out and exercise. My goodness, if you sit all day long, you never get any exercise. And I thought it would be a better job than what I had.

COLLEEN: Who was the chief naturalist who told you not to apply?

RACHEL: His last name was Steenberg, I believe. He was assistant chief. When the second job came open, some of the guides came and talked to me about applying for it.

COLLEEN: Did some of the guides think it was a good idea for you to apply?

RACHEL: They said, "We know we're going to get a woman. It's always been men. It's a men's guide house; everything's been for the one sex—for males only. We wish you would apply and get it. You know us well enough that if we tell an off-color story, you wouldn't get offended." That was one of their excuses for wanting me down there! So I got the job. There wasn't too much said down there that would offend anybody, to tell the truth.

Down in the guide lounge, the front was a two-way mirror. Is it still that way?

COLLEEN: Yes.

RACHEL: They had the couches at the back end toward the restroom. They had a little table up front; they'd sit there to eat. They'd be eating and say, "Hey fellas, there goes a ten!" They'd all get up and run up there. That went on for quite a while. One day I was down there, and I said, "I've got a job I want you to help me do; let's rearrange the furniture." I got them to move all the couches to the front next to that mirror and put the table in the back. They said, "Why do you want to do this?" I said, "So you can all see the number tens go by!"

COLLEEN: Did your male coworkers treat you well?

RACHEL: Most of the time. One incident that I had—I was new. I didn't know where to gather the groups, and I was taking the four-and-a-half-hour trip—back then it was called the Scenic Trip. This day there was about three hundred people. There were three guides, and I was the lead guide; that meant I told the other two what to do. When we got to the dining room, I said, "Fellas, one of you go out front, and when half the group finishes eating, you take them and go on, and I'll follow with the last half. When I got out there, they were still sitting there. We had three hundred people, and there aren't many places on the tour route you can group three hundred people together. My boss, the chief, felt like something was happening. He went in the Frozen Niagara Entrance, met me down the trail, and said, "Where is somebody?" I said, "They didn't split the group like I told them to." That never happened again. There were a few little incidents, but not a lot. We had some men who didn't appreciate me being chief over them.

COLLEEN: When did you become chief guide?

RACHEL: I don't remember the date. I was a regular guide; then they came up with the assistant chief. I applied for that. All I did was put my application in. I was out picking blackberries, and when I got home, I got a phone call saying I got the position. One guide's wife called the chief of interpretation and told his wife that he and I were having a love affair. I was old enough to be his mother! When he got home, she was in bad shape! She was not too happy.

COLLEEN: She thought [the guide's wife] was telling the truth?

RACHEL: The wife didn't know. He got home, and she was all upset. He convinced her, and I said, "Heavenly days!"

It wasn't a happy bunch. A group of them met in Bowling Green with [Congressman] Bill Natcher. They thought something wasn't legitimate with the hiring. I can tell you now, there's no sense in that. But they met with Bill Natcher to see if one of these other fellows could be hired because they had more seniority; they'd been there longer than me. I can see where they'd be upset; I can understand that. The next day the superintendent called me in and said, "Bill Natcher called me. I said, 'I believe the little girl will be alright; don't worry nothing about it.'"

COLLEEN: What did your female coworkers who worked in tickets think of you being a guide?

RACHEL: It didn't matter to them. If they were ever concerned, I didn't know it.

COLLEEN: What about the visitors? Did they say anything?

RACHEL: Not a thing, nothing at all. But when I first got the position, my supervisor would not let me take a trip, so I stayed at the information desk. It went on several weeks. Though I was a guide, I was still at the information desk. One afternoon the superintendent came over and said, "When was the last time you took a cave trip? I said, "I've never taken one." My boss was standing right behind me. He said, "She takes the first one out in the morning." So I took a Frozen Niagara. It had five people on it, and my supervisor went with me. But he didn't let me go for a long time. Their excuse was, "Anything could happen to a woman down there." They were afraid of a single man or two men and a woman guide. He did it, he thought, for my benefit.

COLLEEN: Once you started guiding, and you didn't have the three-inch heels to wear, did you have better shoes?

RACHEL: I had shoes like those [*pointing to my uniform dress shoes*], Oxfords.

COLLEEN: Was the rest of your uniform the same as before?

RACHEL: No, I went into the slacks, just like the men. I got the green slacks, the gray shirt, the Eisenhower jacket, and a parka for the winter time.

COLLEEN: Do you remember what year you started guiding?

RACHEL: No, I don't. It was in the '70s.

COLLEEN: I heard that there was a woman named Willie who was a seasonal guide even before you began guiding.

RACHEL: We had a girl come in; her first name was Barbara—she was the first. She worked one summer. I don't remember where she was; Louis Cutliff [the chief guide] could tell you that information. Willadee Suvee, she was from Alabama. I don't remember that she was there before I was; we must have come in the same season. We got to be pretty good friends. She'd come to my house, and we'd go to church together. She was up here with no relatives.

COLLEEN: I've heard a story that you were throwing torches on a Violet City Lantern Tour and somebody at Chief City threw the torches back at you.

RACHEL: No, that didn't happen to me. Two girls came in to guide. Pat Crowe and Barbara Galooski. They were so lively and full of tricks. Somebody was taking a trip through Chief City, and Pat hid behind a big boulder where they always threw torches. When they threw the torch, she picked it up and threw it back! [*Laughs.*] I never did guide that trip.

COLLEEN: So the women were throwing the torches back to the men?

RACHEL: Yeah.

COLLEEN: What kind of changes did you see from 1957 to 1980?

RACHEL: Every time we got a new superintendent, the first thing they did was change the phone system and the signs. It was so gradual, I wasn't aware. [*That evening, though, Rachel left a phone message for me, saying that the biggest change she'd seen at the park was the increase in female employees.*]

COLLEEN: When you started, the visitor center hadn't been built.

RACHEL: Yes, I began in '57; the visitor center was finished in the early '60s.

One change was that when I began, there was no limit to the number of people they'd take on tours. Anybody who bought a ticket could go on a tour. I saw it gradually change to so many [at a time allowed] down and so many more until it got so there wasn't too awful many. It wasn't unusual to do the Mammoth Dome Trip. Is that what you call it now? Where you go [through] the natural entrance, Fat Man's Misery, River Hall, and up the tower?

COLLEEN: Now we call it Historic Tour.

RACHEL: The Historic Trip when I was there went in the Historic Entrance to the Indian mummy—I mean Giant's Coffin—[and] the TB [tuberculosis] huts and went back out.

The drastic change since I've left is the change in the tickets. It's hideous what people have to pay to go in now. It's a whole lot more. It used to be $1.50 to go on the Historic Trip for adults, children under twelve went free. I believe Frozen Niagara was $1.75; there was a twenty-five-cent bus fee. Last time I heard there was a whole lot of difference.

COLLEEN: Historic Tour is $12 now.

RACHEL: I don't know how families can afford to take their families on the trip. It's too expensive. I don't know why that great of increase in twenty years.

COLLEEN: Do you have any photos of yourself in your old uniform?

RACHEL: Somewhere. I've got a chest of drawers with albums and pictures. I'd have to root through all of them.

COLLEEN: When you first became a guide, did it seem like a big deal that you were one of the first female guides?

RACHEL: I didn't think about it that way. I just had a good job that I enjoyed going to. I never thought about being the first one. And I was really not because Barbara had been there the year before.

I don't know if it's still this way, but they used to hire trailers; they were a lower grade than the guides. They were fours, and the guides were fives. [*GS 4 and GS 5 are government pay grades.*] They'd be there a week before the guide on the trip would "put them up," we called it, on the rock. They started guiding, which wasn't in their job description!

COLLEEN: That's how it was when I started. I was a trailer, but I guided anyway.

RACHEL: That's right! [*Laughs.*] They'd put you right up there!

COLLEEN: Do you have any other stories you would like people to hear about?

RACHEL: All the men treated me well, even though they were resentful of me getting the position above them, especially when I got to be the assistant chief. After I had a talk with one or two of them and things calmed down, they did their job; they didn't let that interfere with their work. Because I got to be above them, I hadn't been there that long, and I was a female—two or three things marked against me, but they were alright.

I never used the bathroom downstairs—never. I always went to the ladies room upstairs. Now the women go in that bathroom like it belonged to them.

COLLEEN: Yes, we all share.

RACHEL: Yeah, I can't see that at all! [*Laughs.*] I spent most of my spare time up at the information desk; I didn't sit down in the guide lounge.

It was a fun place to work. If it wasn't too busy, a lot of fun things went on. You've heard of the puppet show?

COLLEEN: Yes, we still do the puppet show. [*We hold puppets up to a regular window above the one-way glass in the guide lounge for visitors to see.*]

RACHEL: A man came, I think, from Colorado. He surveyed to

put in the shaft for the water system for the Scenic Trip on the top of Mt. McKinley. One of our guides said, "They're going to miss the cave; [the survey,] it's wrong; the cave is over here so many feet." They liked Leo [Hunt, a guide]. I don't think he had a grade-school education; I doubt it very much. He did what we called water witching with a clothes hanger or something, and he could tell where things were. Well, sure enough, they missed the cave. I don't know whether the man came back and surveyed or if they went by Leo's findings, but they redrilled and got the opening right.

18

Nuclear-Fallout Shelters in Mammoth Cave

I found a letter dated November 1962 from the U.S. Army Engineer District to the Edmonson County, Kentucky, Office of Civil Defense that it was working on getting a shelter license to place nuclear-fallout shelters in Mammoth and Great Onyx Caves.[1] The date on the letter didn't surprise me; the Cuban Missile Crisis had occurred that October. Soviet missiles in Cuba aimed at the United States brought the United States and the Soviet Union the closest they ever came to nuclear war. The Cold War prompted the creation of nuclear-fallout shelters throughout the United States, Europe, and the Soviet Union. Most of these shelters were in basements, so the world's longest natural basement became a civil defense fallout shelter between 1963 and 1978.

Nuclear Fallout

When a nuclear bomb explodes, debris is drawn up into a nuclear cloud as high as fifteen miles.[2] This debris becomes contaminated with radioactive material and falls back to earth, giving off gamma rays, most of which are emitted in the first twenty-four hours after the debris has settled. Fallout can return to the immediate area of the explosion within fifteen minutes of the blast and reach areas two hundred miles away in five to ten hours.[3] Exposure to radiation can cause death quickly or have later effects, including leukemia and genetic problems.

Mammoth Cave as a Fallout Shelter

Heavy, dense materials, such as concrete, rock, and earth, offer protection from radiation because they absorb gamma rays.[4] Mammoth Cave has plenty of rock and earth to shield people from radiation. According to Charles H. Bogart of Kentucky Disaster and Emergency Services, the cave would probably be as safe as fallout shelters in basements.[5]

The Snowball Room and Crystal Cave sections of Mammoth Cave and Great Onyx Cave had shelter supplies and probably would have been suitable. Audubon Avenue near the Historic Entrance, which had supplies for the largest number of people, probably would not have made as good a shelter in the winter unless the entrance gate were sealed (the Historic Entrance did have a more restrictive gate when the shelter was in use). Without a solid door, the breeze that blows in the Historic Entrance during cold weather would probably have brought in nuclear fallout.

Frozen Niagara and New Entrance were considered but rejected as sites for shelters.[6]

Caves in General as Shelters

Even though Mammoth Cave might make a suitable fallout shelter, most caves would not be good shelters. Most of them are not in highly populated areas and do not have roads to them or trails inside them, which makes them hard to get to and through. A fallout shelter in a cave passage near water wouldn't work well; the water might wash in radiation, or the passage might be prone to flood. A well-ventilated cave might let in fallout, and a cave with too little ventilation might be unsafe for large groups of people for long periods of time. The cool temperature of most caves in the United States would be uncomfortable for inactive people.[7]

In general, because buildings are created to suit human needs, they make better fallout shelters than caves do.

Shelter Supplies in Mammoth Cave

Audubon Avenue near the Historic Entrance had supplies for 5,000 people, Marion Avenue near the Snowball Room for 500, Crystal Cave for 1,000,[8] and Great Onyx Cave for 1,500.[9] Yellow-and-black fallout-shelter signs at the entrances informed the public of the shelters inside.

The federal government provided supplies for the caves like those in other public shelters throughout the country (although the quantity differed from shelter to shelter): food, water, a medical kit, sanitary supplies, and a radiation kit. The food included a wheat-and-corn-flour survival cracker, which looked similar to a graham cracker,[10] as well as pineapple- and cherry-flavored hard candy for carbohydrates. Public shelters provided ten thousand calories per person, assuming full capacity. How much a person would get for each meal depended on how many people actually arrived at the shelter and how long a stay was anticipated. Both the crackers and the candy were about two thousand calories per pound and had a shelf life of five to fifteen years.[11] The Office of Civil Defense recommended at least three and a half gallons of water per person.[12] Empty seventeen-and-one-half-gallon barrels were shipped to the cave and filled with water on arrival.[13]

Medical kits contained laxatives, petroleum jelly, aspirin, baking soda, bandages, thermometers, penicillin, and phenobarbital, a barbiturate that could be used to control emotional problems caused by the stress of the situation.[14]

Sanitation supplies were stored in a fiberboard drum, which became the commode. The sanitation supply drum held a chemical for use in the commode, a plastic liner, a toilet seat, toilet paper, sanitary pads, hand cleaner, drinking cups, and rubber gloves.[15] The water barrels could also become commodes after the water had been drunk.[16]

Radiation kits contained a Geiger counter, a radiological survey meter, a radiological dosimeter, and a dosimeter charger. These instruments would be used to check radiation levels in the shelter, people, or supplies entering the shelter after fallout began and in adjoining areas to see if they were safe to enter.[17]

The End of the Shelters

Officials at Mammoth Cave National Park had the Youth Conservation Corps and Young Adult Conservation Corps remove the fallout-shelter supplies in 1978.[18] They poured the water out in the caves before removing the barrels. In Audubon Avenue, this water washed away sediment, leaving cracks in the packed dirt.[19] The crackers from many fallout shelters were shipped to Africa and India under the assumption that if people were hungry enough, they would overlook staleness and the bad taste.[20] Some of the crackers from Mammoth Cave were taken to the guide lounge for Park Service employees to eat, but the cave guides weren't that hungry.[21] Local farmers used some of the crackers as hog food.[22] The carbohydrate-supplement candy was a bigger hit with cave guides; several remember eating it, though one said it was "hard as rocks."[23] Workers threw away the first-aid supplies.[24] They weren't always careful while removing the shelter supplies, though; in Crystal Cave, they destroyed an area decorated with gypsum flowers, Nanny Ramsey's Flower Garden.[25]

The shelter supplies in Mammoth and Great Onyx Caves were never replaced. The threat of nuclear war has decreased since the end of the Cold War (though it wasn't yet over when the supplies were removed in 1978). We don't see many of the yellow-and-black fallout-shelter signs that were once common, and the Office of Civil Defense no longer exists. We can only hope that we will never feel the need to restock the caves with such supplies again.

19

The Literary Mammoth

What do Moby-Dick, an Aztec mummy, and Italian robbers have in common? They're all in stories that mention Mammoth Cave.

Both fiction and nonfiction writers have been fascinated with the cave since the early 1800s. Many books and stories feature the cave prominently; others just briefly mention it. Countless newspaper articles and scientific works have also been written about the cave. In my survey of such literature here, I stick to fiction and popular nonfiction aimed at the general public because if I included all nonfiction, the list would be too mammoth.

Modern Kids' Books

Danger at Sand Cave by Candice F. Ransom (Millbrook Press, 2000). Arly Dunbar, a fictional kid, assists with the rescue attempt of the historical figure Floyd Collins, a caver trapped in Sand Cave near Mammoth, by bringing coffee and newspapers to the rescuers. Arly attempts to reach Floyd, but his lantern goes out in the cave. Although Arly manages to make it out alive, (spoiler alert!) Floyd dies. This book is based on the true story of Floyd Collins.

Journey to the Bottomless Pit: The Story of Stephen Bishop & Mammoth Cave by Elizabeth Mitchell, illustrated by Kellyn Al-der (Viking, 2004). Even though we don't know a great deal about early-nineteenth-century mixed-race cave guide Stephen Bishop (1821–1857), he has become a popular character in historic fiction. The book tells of Stephen's experiences as a guide and an explorer.

The Bunyans by Audrey Wood (Scholastic Press, 2006). Giant woman Carrie McIntie digs out Mammoth Cave, where she meets and marries the legendary giant lumberjack Paul Bunyan. Many other tales of the big couples' big adventures are included in the book.

Though the Paul Bunyan legends have roots in nineteenth-century logging camps, this story and others are the products of twentieth- and twenty-first-century writers' imaginations.

Underground by Jean Ferris (Farrar, Straus and Giroux, 2007). *Underground* is written from the point of view of Stephen Bishop's wife, Charlotte, beginning when she first arrives at the cave and meets him. Stephen and Charlotte take great risks helping runaway slaves who stop at Mammoth Cave.

Modern Literature for Young Adults and Adults

The Mammoth Incident by George R. Harker (Dr. Leisure, 1995). Operative Mick Camden from the Special Services branch of the National Park Service is called to break up a drug ring in the Mammoth Cave area. (You didn't know about that branch, did you? That's because it's not real.). The drug search is put on hold when terrorists trap visitors on a cave tour by blowing up passages and hold them hostage for $10 million. Millie Malone, a beautiful young seasonal guide, is one of the few who know an alternate route to the passage where the hostages are held. Can Camden and Millie save lives and $10 million? *Warning:* This book contains cave action beyond crawling.

Ultima Thule by Davis McCombs (Yale University Press, 2000). Cave guide and poet Davis McCombs won the Yale Series of Younger Poets award with this collection of Mammoth Cave poems. One poem about crossing Bottomless Pit is written from Stephen Bishop's perspective:

> Before I crossed it on a cedar pole, legs
> dangling into blackness, here tours
> would end: a loose and shingly precipice.
> from my pack I would produce a scrap
> of oiled paper, set fire to it, and send it
> twisting and sputtering into the abyss.
> I never saw it land, a flicker of light
> on the fluted cistern. Soon I had found
> the rivers beyond, the strange inhabitants
> that emerged into the circle of my light
> as if from another world, then vanished
> at the least agitation of the water. *Touched,*
> they said, *fish with no eyes!* Until I sloshed
> a pailful into light, reveled in their silence.[1]

A Scattering of Jades by Alexander C. Irvine (Tom Doherty Associates, 2002). This historic fiction mixed with fantasy features historic cave guides Stephen Bishop and Mat and Nick Bransford, cave owner Dr. John Croghan, P. T. Barnum, and former vice president Aaron Burr. Throw in an Aztec mummy that contains a spirit called a "chacmool" that Stephen finds in the cave, the possible destruction of humanity, and a difficult decision for Stephen, and you have a weird but fun read.

Beneath Their Feet by Patricia H. Quinlan (iUniverse, 2004). The author is the former wife of the late Jim Quinlan, who used to be a Mammoth Cave hydrologist and avid collector of Mammoth Cave papers, many of which Pat used in writing this novel. *Beneath Their Feet* is historic fiction that tells the story of cave guides from the discovery of the mummified remains of an American Indian in 1811 to the establishment of Mammoth Cave National Park in 1941. The book was self-published, so it may be hard to find.

Grand, Gloomy, and Peculiar by Roger Brucker (Cave Books, 2009). A longtime cave explorer, Roger Brucker is also the author or coauthor of *The Caves Beyond, The Longest Cave, Trapped!,* and *Beyond Mammoth Cave,* nonfiction books about cave exploration. *Grand, Gloomy, and Peculiar* is historic fiction about cave guide and slave Stephen Bishop and his wife, Charlotte, written in the first person with Charlotte as the narrator. The story is based on historic events, but, like most authors of historic fiction, Brucker takes artistic license in plot and character development.

River Runs Deep by Jennifer Bradbury (Atheneum Books for Young Readers, 2015). This historic-fiction novel is about a young patient in the cave's tuberculosis hospital in the 1840s. The author develops personalities for historic characters and creates a fun fictional plot to make a good story while remaining fairly true to the cave's history.

Nineteenth- and Early-Twentieth-Century Fiction

A Wonderful Discovery! An Account of a Recent Exploration of the Celebrated Mammoth Cave of Edmonson County, Kentucky, by Dr. Rowan, Professor Simmons and Others, of Louisville, to Its Termination of the Earth, anonymous (R. H. Elton, 1839). The cave's seemingly endless passages inspired the anonymous author to write of another world deep within the earth's interior that can be reached through Mammoth Cave. This land inside the earth receives sun through holes in the North and South Poles. People who live in the inner earth ride giant birds called "om-mos." The story is in the same vein as *Journey to the Center of the Earth* but predates Jules Verne's famous novel by twenty-five years.

The History of Ester Livingstone and the Dark Career of Henry Baldwin, anonymous (Elmer Barclay, 1853). Ester learns that her husband, Henry, is having an affair with Naomi, one of their slaves. Ester kills Naomi and hides her body behind a brick wall. Henry finds the body and realizes Ester killed Naomi, so he suggests a nice vacation to Mammoth Cave. While on a tour, the couple wanders away from their group and comes to a cliff, where Henry pushes Ester to her death—or so he thinks.

After more adventures, Henry returns home to his mother and sisters. While alone in his room one night, though, he finds Ester, very much alive, at the window. She had fallen only a short distance in the cave, survived, and found her way out. After selling the diamonds she was wearing, she had enough money to follow Henry and watch him as he went on with his life. Henry then takes a dagger, kills her, drags her body out the window, and throws her into a lake.

"A Tragedy of the Mammoth Cave" by Lillie Devereux Blake (*The Knickerbocker,* 1858). Widowed in her twenties, Lillie Devereux Blake supported herself by writing magazine articles, short stories, and novels. A women's rights advocate, she was also a war correspondent during the Civil War, an unusual position for a woman at the time.

The Knickerbocker, a popular literary magazine in the mid-nineteenth century, featured Blake's short story "A Tragedy of the Mammoth Cave," a tale of a young girl named Melissa in love with her tutor, William, who does not return her affection. While Melissa sits in Mammoth Cave near Echo River pining for William, a tour arrives to cross the river; William and a female companion are on it. There is no room on the boat for William, so the guide says he will return for him and leaves with the group. To William's surprise, Melissa appears from behind the rocks and offers to take him on an alternate route to the other side of the river. As they walk through the cave, she asks him about the girl. When he says they are engaged, Melissa blows out her lantern, slips away (she knows the way in the dark), and leaves him alone in the darkness. Her plan is just to scare him, but William is never found again. Years later, guilt drives her to seek the same fate for herself: "I am going to reenter that dark Cave, the threshold of which I have not crossed for fifteen years, and there I will patiently await the coming of that death, which I hope to me will be a blessed release. The gloom and horror of which, years ago, I doomed my victim, shall be around me when I die: for I think that perhaps from amid the silent rocks which witnessed my crime, my last prayer for forgiveness will find acceptance."[2]

In the twentieth century, some people reported hearing the ghosts of Melissa and William in the cave. They apparently didn't know the story is fiction!

"Legend of Mammoth Cave," in *Legends of the South* by Nathan Ryno Smith (Steam Press of William K. Boyle, 1869). A traveler hears about an American Indian who years earlier visited Mammoth Cave once a year to spend twenty-four hours underground. When the Indian was a child, his people were surrounded by white men in battle. Rather than die at enemy hands, the Indians entered the cave, where they all perished, leaving the child to be the last of his tribe. The traveler learns that upon the Indian's last visit, he said, "This is the last visit that I shall ever make to the tomb of my tribe. I shall enter and you will see me no more." The curious traveler enters the cave and meets the spirit of the Indian, who predicts the consequences of the white man's evil deeds against the Indians: "The bear of the North shall wage war with the panther of the South, and terrible shall be the conflict,"[3] forecasting the Civil War.

Mammoth Cave Romance by William Lee Popham (Mayes Printing, 1911; reprint, Byron's Graphic Arts, 1993). *Mammoth Cave Romance* contains more romance than cave. Here's the gist: boy meets girl at Mammoth Cave; boy loses girl; boy meets same girl by chance at the cave again; boy gets girl back; and they live happily ever after.

"The Beast in the Cave" by H. P. Lovecraft (*The Vagrant,* June 1918; reprinted in H. P. Lovecraft, *Dagon and Other Macabre Tales* [Arkhan House, 1965]). H. P. Lovecraft was a popular writer of macabre stories. This was one of his earliest works, written in 1905, when he was fourteen years old, but published later. The tale takes advantage of three common cave fears—being caught in total darkness, getting lost, and coming across dangerous animals that lurk in the dark. The main character sees the worst possible result of wandering away from your tour at Mammoth Cave when he gets lost and meets a scary cave beast. He mortally wounds the creature, takes a closer look, and sees what could be his own fate if he's not rescued: "The creature I had killed, the strange beast of the unfathomed cave was, or had at one time been, a MAN!!!"[4]

Fiction That Briefly Mentions Mammoth Cave

Autobiography of a Pocket-Handkerchief by James Fenimore Cooper (Mythik Press, 1843). This social satire about greed and materialism is told from the unusual perspective of a handkerchief. Cooper, best known for his novel *The Last of the Mohicans* (1826), mentions unnamed Kentucky caves in this less-well-known work. Mammoth was most likely his inspiration because few caves were open to the public in 1843: "You once raised me from the depths of despair to an elevation of happiness that was high as the highest pinnacle of the caverns of Kentucky; raising me from the depths of Chimborazo."[5]

Cooper apparently had some depth-height confusion; Chimborazo is the highest mountain in Ecuador.

Moby-Dick by Herman Melville (Harper and Brothers, 1851). In his most famous novel, Melville compares a dead sperm whale to Mammoth Cave. "Let us now with whatever levers and steam-engines we have at hand cant over the sperm whale's head, so that it may lie bottom up; then, ascending by a ladder to the summit, have a peep down the mouth; and were it not that the body is now completely separated from it, with a lantern we might descend into the great Kentucky Mammoth Cave of his stomach."[6]

According to the Local History Department at Berkshire Atheneaum in Pittsfield, Massachusetts, there is no record of Melville visiting either Mammoth Cave or Kentucky. Perhaps he knew the cave only by its reputation. He apparently had an interest in it because he also mentioned it in two subsequent though less-famous novels.

Mardi and a Voyage Thither by Herman Melville (Harper and Brothers, 1849). Like *Moby-Dick,* this Polynesian adventure was influenced by Melville's interest in the sea. To point out that all great things take time to make, he mentions Mammoth Cave in a long list of wonders, natural and man-made. "Nero's House of Gold was not raised in a day; nor the Mexican House of the Sun; nor the Alhambra . . . nor Stonehenge; nor Solomon's Temple. . . . No: nor were the great grottos of Elephanta hewn out in an hour; nor did the Troglodytes dig Kentucky's Mammoth Cave in a sun."[7]

The Confidence-Man by Herman Melville (Dix, Edwards, 1857). Melville knew about the tuberculosis hospital in the cave in the 1840s. In his last novel, *The Confidence-Man,* the main character talks to a man with a bad cough.

> But, while descending a stairway, he was seized with such coughing that he was fain to pause.
>
> "That is a very bad cough. . . . Have you tried anything for it?"

> "Tired of trying. Nothing does me any good—ugh! ugh! Not even the Mammoth Cave. Ugh! ugh! Denned there six months, but coughed so bad the rest of the coughers—ugh! ugh!—black-balled me out. Ugh, ugh! Nothing does me good."[8]

Journey to the Center of the Earth by Jules Verne (Hetzel, 1864). The French science fiction writer Jules Verne, like Melville, was intrigued with Mammoth Cave. Verne mentioned it in several works, but his fascination was based on what he read or heard about the cave rather than what he would have seen in a visit; his only journey to the United States was a short trip to New York.

In *Journey to the Center of the Earth,* the characters descend down a volcano in Iceland into the interior of the earth, which is a world in itself. Harry, the narrating character, compares the center of the earth to famous caves. "I had read of most wonderful and gigantic caverns—but none in any way like this. The great grotto of Guachara, in Columbia, visited by the learned Humboldt; the vast and partially explored Mammoth Cave in Kentucky; what were these holes in the earth to that in which I stood in speechless admiration!"[9]

All Around the Moon by Jules Verne (Hetzel, 1870). This book is also known as *Around the World* and *Round the World.* Long before space travel became a reality, Verne's three astronauts circle the moon, viewing craters and looking for life. "They looked out of the windows once more at the black moon beneath them. There it lay below them, a round black spot, hiding the sweet faces of the stars, but otherwise no more distinguishable by the travelers than if they were lying in the depths of the Mammoth Cave of Kentucky."[10]

Their space ship eventually returns to earth and lands in the ocean, where the astronauts are rescued by the U.S. Navy ship *Susquehanna.*

The Mysterious Island by Jules Verne (Hetzel, 1874). During the Civil War, five Union prisoners of war and a dog escape from the Confederates by stealing a hot-air balloon. They land on an uncharted island, where things go remarkably well for them. They find a box of guns and tools; the dog is mysteriously rescued from a wild beast; they are lost in a storm while sailing their homemade boat but find their way back to the island by following a flare that none of them lit; and when pirates attack, the pirates' ship mysteriously blows up. The castaways eventually find the secret to their mysterious good luck in a cave. "Such immense caves exist in various parts of the world, natural crypts dating from the geological epoch of the globe. Some are filled by the sea; others contain entire lakes in their sides. . . . [S]uch are the immense Mammoth caverns in Kentucky, 500 feet in height, and more than twenty miles in length! In many parts of the globe, nature has excavated these caverns, and preserved them for the admiration of man."[11]

Captain Nemo, captain of the submarine the *Nautilus* in Verne's earlier novel *Twenty Thousand Leagues under the Sea* (1870), is their secret caretaker. Nemo keeps his sub in a port in the cave, which is accessible from the sea. He had retired to the island after everyone else on the *Nautilus* had died.

Captain Nemo eventually dies and is buried at sea in his submarine, the island blows up, and the castaways are rescued by a ship.

The Underground City by Jules Verne (Hetzel, 1877). Also called *The Child of the Cavern,* this tale is about a city 1,500 feet under Scotland. Verne mentions Mammoth Cave in two places: "If, by some superhuman power, engineers could have raised in a block, a thousand feet thick, all that portion of the terrestrial crust which supports the lakes, rivers, gulfs, and territories of the counties of Stirling, Dumbarton, and Renfrew, they would have found, under that enormous lid, an immense excavation, to which but one other in the world can be compared—the celebrated Mammoth Caves of Kentucky"; and "Below the dome lay a lake of an extent to be compared to the Dead Sea of the Mammoth Caves—a deep lake whose transparent waters swarmed with eyeless fish, and to which the engineer gave the name of Loch Malcolm."[12]

Facing the Flag by Jules Verne (F. T. Neely, 1894). This novel may have inspired later books and movies about power-hungry mad scientists and bad guys who have secret island lairs.

A crazy French scientist named Roch invents the Fulgurator, a superweapon so powerful that "the state which acquired it would become absolute master of earth and ocean."[13] He tries to sell the Fulgurator to the French, German, British, and American governments, but because there is no proof the Fulgurator works, they're not interested.

A pirate named Ker Karraje hears about the Fulgurator and decides a good superweapon would be useful to protect his hideout in a cave (accessible only by submarine) on the tropical island of Back Cup. Verne compares the Back Cup cave to Mammoth.

> The size of the cavern can be judged from these approximate figures. But vast as it is, I remember that there are caverns of larger dimensions both in the old and new worlds. . . . [T]hose at Han-sur-Lesse in Belgium, and the Mammoth Caves in Kentucky, are also more extensive. The latter contains no fewer than two hundred and twenty-six domes, seven rivers, eight cataracts, thirty-two wells of unknown depth, and an immense lake which extends over six or seven leagues, the limit of which has never been reached by explorers.
>
> I know these Kentucky grottoes, having visited them, as many thousands of tourists have done. The principal one will serve a comparison to Back Cup. The roof of the former, like that of the latter, is supported by pillars of

> various lengths, which give it the appearance of a Gothic cathedral, with naves and aisles, though it lacks the architectural regularity of a religious edifice. The only difference is that whereas the roof of the Kentucky grotto is over four hundred feet high, that of Back Cup is not above two hundred and twenty at that part of it where the round hole through which issue the smoke and flames is situated.[14]

Karraje kidnaps Roch and his attendant, Gaydon, and takes them to Back Cup. Unbeknownst to Roch, Gaydon is really Simon Hart, a French engineer who is trying to learn the secret of the Fulgurator for the French government. Roch seems to think he is a guest of the pirates rather than a prisoner, but Hart knows what is going on. He puts a message in a keg and throws it into the sea, hoping it will float to British authorities in the Bahamas, which of course it does.

A British submarine arrives at Back Cup, but the pirate sub blows it up. Warships from several countries soon arrive. Roch blows up the first ship with his Fulgurator, but as the next ship approaches, he sees it is flying the French flag. Overcome with patriotism, he does not blow up the ship. When the pirates try to take control of the Fulgurator, Roch blows up the whole island—cave, pirates, and all.

Will of an Eccentric by Jules Verne (Hetzel, 1900). Verne includes a whole chapter about Mammoth Cave in this story about six characters who use the United States as their game board, in which a roll of the dice sends them to sites across the country. The dice send players Lizzie Wag and Jovita Foley to Mammoth Cave—a much desired space on the board because it is "one of the wonders of the United States and, I believe, the world." The girls spend several days taking vigorous cave tours, but it is worth it. "One would generously pay with fatigue to traverse these marvelous caves—a walk through the enchanted world of the Arabian Nights—even without meeting with demons or gnomes, and Jovita Foley was only too pleased to admit that the spectacle exceeded the limits of human imagination."[15]

Originally published in France in 1900, this novel was reprinted in the United States in 2009.

Etidorhpa by John Uri Lloyd (self-published, 1897). Lloyd was a pharmacist well known for his work with medicinal plants. Along with several pharmaceutical books, he wrote fiction. His best-known novel, *Etidorhpa* ("Aphrodite" spelled backward) is about a man known only as "I Am the Man" who descends into another world with giant mushrooms and prehistoric animals in the earth's hollow interior as punishment for giving away the secrets of a mysterious fraternal organization.

The entrance to the inner earth is an unnamed Kentucky cave that is entered by wading into water—apparently not Mammoth Cave. As the Man and his guide journey to the underground world, the guide compares that world to Mammoth Cave: "The Mammoth Cave as now traversed is simply a superficial series of grottoes and passages overlying the deeper cavern field that I have described. The explored chain of passages is of great interest to men, it is true, but of minor importance compared to others yet unknown. . . . Echo River, is a miniature stream: there are others more magnificent that flow majestically far, far beneath it."[16]

Some people believe *Etidorhpa* is more than science fiction. Joseph H. Cater says in his book *The Ultimate Reality* that "some books are written in the form of a novel in order to present certain ideas or truths without inviting undue attack from various quarters. *Etidorhpa* is considered by most to be a science fiction book. Any intelligent and discerning reader realizes that it isn't."[17]

"The Box of Robbers," in *American Fairy Tales* by L. Frank Baum (George M. Hill, 1901). L. Frank Baum, best known for his Oz series, starting with the classic *The Wonderful Wizard of Oz* (1900), wrote several other children's books, including a collection of short stories titled *American Fairy Tales.* In the first tale, "The Box of Robbers," young Martha unlocks a mysterious chest her uncle Walter has sent from Italy. To her surprise, three robbers jump out. Upon learning their profession, Martha tries to convince the robbers to take up another line of work. They insist that they are suited to no other occupation and begin to rob Martha's family's house.

> All bore heavy loads of plunder in their arms, and Lugui was balancing a mince pie on the top of a pile of her mother's best evening dresses. Victor came next with an armful of bric-a-brac, a brass candelabra and the parlor clock. Beni had the family bible, the basket of silverware from the sideboard, a copper kettle and papa's fur overcoat.
>
> "Oh joy!" said Victor, putting down his load; "it is pleasant to rob once more."
>
> . . . "We have much wealth," continued Victor, holding the mince pie while Lugui added his spoils to the heap: "and all from one house! This America must be a rich place."
>
> . . . "We should have a cave," remarked Beni; "for we must store our plunder in a safe place. Can you tell us of a secret cave?" he asked Martha.
>
> "There's a Mammoth Cave," she answered, "but it's in Kentucky. You would be obliged to ride on the cars a long time to get there."[18]

Later, when the postman rings the doorbell, Martha tells the robbers the police are at the door. The robbers panic about how they will escape from the police, so Martha offers to hide them in the trunk. They jump back in their box, where Martha safely locks them in.

An Undivided Union by Edward Stratemeyer (Lee & Shepard, 1899). This is the last book in the author's Civil War series "Blue and Gray on Land." Union cavalrymen are looking for Confederate soldiers when their way is blocked by a waterfall and pool.

> "Here are horses' hoof-prints, Major," said one of the men. "I shouldn't wonder if there is a winding path leading down to that 'air pool. But if the rebs went down there, what became of 'em?"
>
> "There may be a cave there," answered Deck. "These underground watercourses often flow through caves around where I live, not far from the Mammoth Cave."[19]

Keeping Up with Lizzie by Irving Bacheller and William Harlowe (Harper & Brothers, 1911). This novel about social climbing was made into a movie in 1921. The cave is used as a metaphor in describing one of the participants in a discussion about cleaning one's own house versus having hired help. "'I think it's terrible,' said a fat lady from Louisville, distinguished for her appetite, and often surreptitiously referred to as 'The Mammoth Cave of Kentucky.' 'The idea of trying to make it fashionable to endure drudgery! I think we women have all we can do now.'"[20]

Nineteenth-Century Nonfiction

Notes on the Mammoth Cave, to Accompany a Map by Edmund F. Lee (James & Gazlay, 1835). In addition to his book describing points of interest in the cave, Lee surveyed the cave and made a map called the "Lee Map." He determined that the cave was about eight miles long. Lee's map and guidebook became outdated three years later when the Bottomless Pit was crossed, adding many more miles and sights to see.

Peter Pilgrim: or, A Rambler's Recollections by Robert Montgomery Bird (Lea & Blanchard, 1838). Bird was a medical doc-

Nineteenth-century writers toured the cave with groups like this one. (Courtesy National Park Service)

tor turned poet, playwright, novelist, and artist. *Peter Pilgrim* includes a story that Bird's guide told him on a Mammoth Cave tour. The story, titled "The Miner and the Devils," tells about the lost saltpeter miner whose light went out in Gothic Avenue in Mammoth Cave.

> [H]e felt himself a doomed man, he thought his terrible situation was a judgment imposed upon him for his wickedness; nay, he even believed, at last, that he was no longer an inhabitant of the earth . . . in other words, that he was in hell. . . . It was at this moment the miners in search of him make their appearance: . . . all swinging their torches aloft, he[,] not doubting they were those identical devils whose appearance he had been expecting, took to his heels, yelling lustily for mercy; nor did he stop, notwithstanding the calls of his amazed friends, until he had fallen a second time among the rocks, where he lay on his

> face, roaring for pity until by dint of much pulling and shaking, he was convinced that he was still in the world and the Mammoth Cave.[21]

Cave guides still entertain and scare visitors with the miner's story, but Bird's subsequent testimonial will make you forget your fear. "I recommend all broken-hearted lovers and dyspeptic dandies to carry their complaints to the Mammoth Cave, where they will undoubtedly find themselves 'translated' into very buxom and happy persons, before they are aware of it."[22]

Bird also did a painting of the Mammoth Cave entrance and one of the buildings, which he called *The Old Cabin (Gatewood's House) at the Mammoth Cave.*

Letters from New York by Lydia Maria Child (Francis, 1845). In her day, Child was famous for writing about the abolition of slavery (an unpopular stand that made her work sell less) as well as about women's and Native Americans' rights. Today, her most remembered work is the children's poem and song "Over the River and through the Woods to Grandfather's House We Go."

Though Child herself did not visit Mammoth Cave, a friend's report of his visit inspired her to write a detailed ten-page description of the cave in her book *Letters from New York.* She wrote of the cave guide and slave Stephen Bishop as if she had met him. "Stephen, the residing genius of Mammoth Cave, is a mulatto, and a slave. He has lived in this strange region from boyhood, and a large proportion of the discoveries are the result of his courage, intelligence, and untiring zeal. . . . His knowledge of the place is ample and accurate, and he is altogether an extremely useful and agreeable guide. May his last breath be a free one!"[23]

Rambles in the Mammoth Cave during the Year 1844 by a Visiter [*sic*] by Alexander Clark Bullitt (Morton & Griswold, 1845; reprint, Johnson Reprint Corporation, 1973). It is clear Bullitt read *Notes on Mammoth Cave, Peter Pilgrim,* and *Letters from New York* because he plagiarized sections of these books in *Rambles.*

A cave map drawn by cave guide Stephen Bishop was published in *Rambles.* Interestingly, though Bishop was a slave, he was openly given credit for his map in the book, when the author himself didn't receive such credit in the 1844 edition, which identifies him as only "a Visiter."

The book's republication in 1973 made it popular with modern Mammoth Cave fans.

The Sucker's Visit to the Mammoth Cave by Ralph Seymour Thompson (Live Patron Publishing Office, 1879; reprint, Johnson Reprint Corporation, 1970). Shortly after the Civil War, Thompson and four companions set out on an old-fashioned (even by standards in the 1860s) adventure from Illinois to Mammoth Cave. His description of a cave tour includes lunch with guide Mat Bransford in the cave.

> Near the middle is a large flat rock, and on this Matt [*sic*] set down the huge basket he had been so patiently carrying, and proceeded to unpack a very excellent dinner.
>
> But we had to take exceptions to the serving of the dinner. It is romantic to sit on rocks and eat your dinner with your fingers, but it is much more convenient to sit on chairs and eat it in the good old way. And there is no reason in the world why this grand natural dining hall should not be furnished with tables and comfortable chairs, so that visitors could really rest while at dinner; and no very serious reason why knives and forks and plates should not be provided.[24]

Seymour describes their misadventures to and from the cave, so you get plenty of good travel stories in addition to the tour description typical of historic accounts of the cave. According to Seymour, the group found more "romance" than they planned by taking deck passage on a riverboat in their trip home.

> But what about deck passage? Well, the result of my experience might be summed up in one word of advice to any other young man who has any idea of taking deck passage for the romance of the thing:
>
> DON'T!
>
> If being kicked and cuffed about like a dog is romantic, then deck passage is romantic; yes, extremely so. . . .
>
> If having to take your chances for sleep on boxes, barrels, and coils of rope, and be regularly ordered—by some deck hand—to "get off that," as soon as you get partly asleep, is romantic, then deck passage is romantic; yea, a very epitome of romance. . . .
>
> Yes, if all these things are romance, then this trip was the most romantic episode of our lives. But I can't stand so much romance—not of this kind.[25]

One Hundred Miles in Mammoth Cave in 1880 by Horace C. Hovey (Scribner's Monthly, October 1880; reprint, Outbooks, 1982); *Guide Book to the Mammoth Cave of Kentucky* by Horace C. Hovey (Clark, 1895); *Celebrated American Caverns, Especially Mammoth, Wyandot, and Luray: Together with Historical, Scientific, and Descriptive Notices of Caves and Grottoes in Other Lands* by Horace C. Hovey (R. Clarke, 1896); and *Mammoth Cave of Kentucky; An Illustrated Manual* by Horace C. Hovey, with Richard Ellsworth Call (John P. Morton, 1912).

Horace Hovey was a Presbyterian minister and geologist who wrote often about caves, especially Mammoth Cave. His books include descriptions of the tour routes and cave landmarks popular with other early writers, but he covered the cave more thoroughly than most writers. Being a scientist, he extensively described the cave's geology and biology. In the revised edition of *Mammoth Cave of Kentucky* (1912), his background as both a scientist and a minister comes through:

> The best definition of evolution describes it as a continual differentiation of the complex to the simple. First, simple forms; then the complex. But in cave fauna we find the process reversed: the complex forms are reverting to those that are more simple. Our limits forbid our either following further such fascinating problems, or taxing the reader's patience by moralizing. Yet we may affirm anew our cherished faith that all forms of life exist and go on under a Divine plan, whether by progression or retardation, by deprivation or compensation, by evolution or devolution, environed by darkness or light, amid profound caverns or amid the brave sunshine. Many things beyond our immediate comprehension are worthy of patient and prolonged investigation.[26]

Nineteenth-Century Nonfiction That Briefly Mentions Mammoth Cave

Health Trip to the Tropics by Nathaniel Parker Willis (Charles Scribner, 1853). A magazine writer and editor, Willis founded and edited the magazine *National Press,* which is still published today as *Town & Country.* He made Mammoth Cave sound colorful and spectacular in his book *Health Trip to the Tropics* when he wrote, "Why, the state apartments of Versailles are not half so sumptuously ornamented as this portion of the basement story of Kentucky."[27]

Walden by Henry David Thoreau (Ticknor and Fields, 1854). Required reading for many high school English students, *Walden* is about Thoreau's two-year retreat to a cabin to contemplate and simplify life. In the chapter "Where I Lived, and What I Lived For," he writes, "Why should we live with such hurry and waste of life? . . . After a night's sleep the news is as indispensable as the breakfast. 'Pray tell me anything new that has happened to a man anywhere on this globe'—and he reads it over his coffee and rolls, that a man has had his eyes gouged out this morning on the Wachito River; never dreaming the while that he lives in the dark unfathomed mammoth cave [*sic*] of this world, and has but the rudiment of an eye himself."[28]

At Home and Abroad by Bayard Taylor (Putnam, 1855). Taylor, a popular travel writer in the mid-1800s, wrote about his adventures in the United States and Europe. Mammoth Cave was closer to home than many of the places he wrote about, but his description makes it sound other-worldly.

> For in the cave you forget that there is an outer world somewhere above you. The hours have no meaning: time ceases to be: no thought of labor, no sense of responsibility, no twinge of conscience, intrudes to suggest the existence you have left. You walk in some limbo beyond the confines of actual life, yet no nearer the world of spirits. For my part I could not shake off the impression that I was wandering on the *outside* of Uranus or Neptune, or some planet still more deeply buried in the frontier darkness of our solar system.[29]

"Illusions," by Ralph Waldo Emerson (*Atlantic,* November 1857), reprinted in *The Conduct of Life* (Ticknor and Fields, 1860). Guide Stephen Bishop took Emerson into Mammoth Cave in 1850, and Emerson wrote an extensive letter to his wife, Lydia, about Bishop, the cave's landmarks, and the illusion of the night sky at Star Chamber. The cave, especially Star Chamber, apparently impressed Emerson because ten years later he described his cave experience at the beginning of his essay "Illusions":

Star Chamber. (Courtesy National Park Service)

But I then took notice, and still chiefly remember, that the best thing which the cave had to offer was an illusion. On arriving at what is called the "Star-Chamber," our lamps were taken from us by the guide, and extinguished or put aside, and, on looking upwards, I saw or seemed to see the night heaven thick with stars glimmering more or less brightly over our heads, and even what seemed a comet flaming among them. All the party were touched with astonishment and pleasure. . . . I sat down on the rocky floor to enjoy the serene picture. Some crystal specks in the black ceiling high overhead, reflecting the light of a half-hid lamp, yielded this magnificent effect.

I own, I did not like the cave so well for eking out its sublimities with this theatrical trick. But I have had many experiences like it, before and since; and we must be content to be pleased without too curiously analyzing the occasions.[30]

The essay is not about the cave but rather about all of life's illusions.

A Thousand-Mile Walk to the Gulf by John Muir (Houghton, Mifflin, 1916). Nineteenth-century conservationist, writer, and Sierra Club founder John Muir is best known for his work to protect the natural wonders of the western United States, especially Yosemite National Park, but he also appreciated and wrote about natural beauty in the East. In 1867, Muir stopped at Mammoth Cave on his walk from Indianapolis, Indiana, to Cedar Key, Florida. He recorded his trek in his book *A Thousand-Mile Walk to the Gulf.* As he approached the cave, he talked to a local man about it:

> He told me that he had never been at Mammoth Cave—that it was not worth going ten miles to see, as it was nothing but a hole in the ground, and I found that his was no rare case. He was one of the useful, practical men—too wise to waste precious time with weeds, caves, fossils, or anything else that he could not eat.
>
> Arrived at the great Mammoth Cave. I was surprised to find it in so complete naturalness. A large hotel with fine walks and gardens is near it. But fortunately the cave has been unimproved, and were it not for the narrow trail that leads down the glen to its door, one would not know that it had been visited. . . .
>
> I never before saw Nature's grandeur in so abrupt contrast with paltry artificial gardens. The fashionable hotel grounds are in exact parlor taste, with many a beautiful plant cultivated to deformity, and arranged in strict geometrical beds, the whole pretty affair a laborious failure side by side with Divine beauty.[31]

"A Tour in the Mammoth Cave," anonymous, in *All the Year Round,* January 19, 1861. This magazine article has no byline, but because Charles Dickens edited *All the Year Round,* some assume he wrote it. According to Jon Michael Varese of the University of California Dickens Project and Robert Newsom of the University of California Department of English and Comparative Literature, however, most of what was published in *All the Year Round* was anonymous, and Dickens probably did *not* write this article. Dickens visited the United States in 1842, but in his journal he wrote that he spent just one night at the Galt Hotel in Louisville while traveling from Cincinnati to St. Louis. He made no mention of the cave or of other sites in Kentucky.

20

Celebrities Underground

Actors, musicians, politicians, and other celebrities come to see the longest cave in the world. Some have arrived with great fanfare, as Ronald Reagan did in 1984; others have blended in so well they would have gone unnoticed if it weren't for observant, nosy cave guides.

Modern Celebrities

Longtime cave guide Keven Neff shared this story about **John Wayne**:

> Shorty Coats, Albert A. Coats [a cave guide], pointed out a rock in the Frozen Niagara section that had John Wayne's name written on it. I said to him, "A lot of people could be named John Wayne, Shorty." He said, "I'm the one that let him write it on there."
>
> John Wayne was just becoming a major actor; I'm not sure what year it was. The date is impossible to read, the name is disintegrating quite a bit. If you hold your light just right, you can see it. But Shorty said he was there when John Wayne did it. But that was before it was a national park. Whenever he was using his stage name, John Wayne. But I believe Shorty that he was here.

Keven also saw **Elton John**: "I was sitting in the guide

house, and Elton John came walking down the ramp. He had on his big fat hat, loud shirt, and sunglasses. Someone said, 'There's Elton John; he's incognito, so no one would see him.' He stood out like a sore thumb!"

The apparently ever-present Keven wasn't there to see **Johnny Cash**, but of course he heard the story:

> I know Johnny Cash came here; chief interpreter Lewis Cutliff took him in. He was in and out of the cave before the regular visiting hours. They took him in real early in the morning. I said, "That's not fair; he should have bought a ticket like everybody else." But then you get to thinking, if you had somebody as well known as Johnny Cash on your group, and you're trying to talk to people, you wouldn't have many people paying attention to you; they'd be looking over at the celebrity. So I understand why celebrities are taken in separately. Not because they're something special, but [because] they would disrupt everything else.

Keven's not the only one with stories about brushes with greatness, though. While working in ticket sales in the spring of 1994, I saw a woman who looked like **Karen Grassle**, the actress who played Caroline Ingalls on the TV show *Little House on the Prairie* (1974–1983), buy a cave tour ticket. When she left, a coworker told me he also thought she looked like Karen Grassle. We looked at the woman's credit card bill, and, sure enough, it was her. I guided her tour the next day. Cave visitors apparently didn't recognize her.

In the summer of 2009, my husband, Rick, and I were eating at a local pizza place while waiting for a play at the theater to begin. I said hello to the theater director, who happened to be there eating with his friends. A big storm rolled in, so instead of going to the play, we went home.

The next day I saw a man who looked familiar in the Mammoth Cave visitor center. "How do I know that man?" I asked myself.

The mysterious familiar man came on my cave tour and asked, "How did you like the play last night?" I thought that's how I knew him; I'd seen him at the pizza place and maybe elsewhere around town. I told him I didn't make it to the play. He said it was great and that he was an old friend of the theater director.

At the end of the tour, a visitor said, "I saw you talking to that man from *Spiderman.*"

"What do you mean?" I said.

"I knew I recognized that man," said the other guide on the tour.

I realized that I recognized the man not only from the pizza place but also from *Spider-Man 2* (2004) and *Spider-Man 3* (2006): he was **Dylan Baker,** who played Dr. Curt Connors, Peter Parker's college professor.

According to another cave guide, **Neil Patrick Harris**, the star of the TV shows *Doogie Houser M.D.* (1989–1993) and *How I Met Your Mother* (2005–2014), took the self-guided tour with a friend sometime around 2005. The cave guide stationed in the cave when Harris came through told me, "He walked briskly past me while asking if the saltpeter vats were left by the Indians. Since he never broke stride, by the time he got the question out, he was already past me, and I hollered, 'No,' to the back of his head."[1]

Some of the celebrities who have passed through the cave are not so well known anymore, though.

- The inscription "Irene Ryan 1937" at the intersection of Boone Avenue and Rose's Pass is one of the few famous names written in the cave. The mystery is whether this **Irene Ryan** was the same one who played Granny on the TV show *The Beverly Hillbillies* in the 1960s. Irene the actress was born in 1902 or 1903 (depending on which source you read), so she was about thirty-five years old in 1937. She performed on the vaudeville stage and radio but wasn't nationally famous until *The Beverly Hillbillies* began airing in 1962 and thus would not have been recognized in her visit to the cave.

- **June Lockhart**, an actress in the 1950s and 1960s TV shows *Lassie* (1954–1974), *Petticoat Junction* (1963–1970), and *Lost in Space* (1965–1968), visited the cave with her daughter in the late 1970s.[2]
- Members of the British band **Supertramp**, famous for their album *Breakfast in America* (1979) and the song "Take the Long Way Home," took a cave tour in 1979.[3]
- **Jim Varney**, best known for his character Ernest in numerous commercials and movies, including *Ernest Goes to Camp* (1987), and for playing Jed Clampet in the movie *The Beverly Hillbillies* (1993), came to the cave with a lady friend in the early 1990s.[4]
- Cave employee by day, pro-wrestling fan by night, Johnny Merideth saw a man he thought looked like wrestler **Jim Cornette** at the visitor center bookstore around 2010. Johnny started to walk toward the man, when a visitor stopped him to ask a question. He considered telling the visitor, "There's a man who beats people with tennis rackets; I have to meet him!" but instead stopped to talk to the visitor and let the wrestler get away. Johnny asked the bookstore cashier if the man bought anything with a credit card. He had, and the name on the card was indeed Jim Cornette.
- In 1999, a big woman had trouble keeping up with her cave tour group, so Alan Sizemore, the guide whose job it was to bring up the rear, slowly walked her out separately. With time to chat, Alan learned she had recently lost a lot of weight and was very spiritual. When the woman's son urged her to tell Alan who she was, she revealed she was **Lulu Roman** from the TV show *Hee Haw*,[5] the country music/comedy show that ran from the late 1960s through the late 1990s, featuring country and bluegrass numbers, corny skits, and scantily clad women.

We had one visitor who was a celebrity only in her own mind. In the mid-1990s, a young woman arrived at the visitor center demanding a complimentary ticket because she starred in a daytime soap opera. Because the guides and ticket salespeople didn't watch soaps, they didn't recognize her and didn't comp her ticket, much to her displeasure. If you are a celebrity and want special treatment, it helps if workers recognize you.

Historic Celebrities

Ole Bull

Norwegian violinist Ole Bull (1810–1880) performed in a room later named Ole Bull's Concert Hall in the spring of 1845. George Prentice, the editor of the *Louisville Courier-Journal,* was also on the tour. He wrote, "Ole Bull took his violin into the cave and gave us some of his noblest performances at the points most remarkable for their wonderful echoes. The music was like no earthly music. It seemed, indeed, superhuman. The whole company were as mute and motionless as statues, and tears, copious and gushing tears, streamed from every eye."[6]

Not everyone liked Bull's playing. The website Violinman.com calls Bull's technique "amateurish" and says his style pleased only people "in localities rarely visited by real artists."[7] I suppose Mammoth Cave in the mid-1800s would fit this description.

In 1853, Nathanial Parker Willis wrote about visiting Ole Bull's Concert Hall with Stephen Bishop. Bishop told him that Bull, George Prentice, and Prentice's wife were the only people on the tour the day Ole Bull played.[8]

Ole Bull may not have been the greatest violinist, but he did get a state park in Pennsylvania named after him, commemorating the place where he tried (and failed) to establish a settlement for Norwegian immigrants.

Ralph Waldo Emerson

Friends of the famous writer and lecturer Ralph Waldo Emerson convinced him to visit Mammoth Cave in 1850. The cave made a lasting impression on him; Emerson began his essay "Illusions" (1857) with a description of a popular illusion at Mammoth Cave.

> Some years ago, in company with an agreeable party, I spent a long summer day in exploring the Mammoth Cave in Kentucky.
>
> . . . But I then took notice and still chiefly remember that the best thing which the cave had to offer was an illusion. On arriving at what is called the "Star-Chamber," our lamps were taken from us by the guide, and extinguished or put aside, and, on looking upwards, I saw or seemed to see the night heaven thick with stars glimmering more or less brightly over our heads, and even what seemed a comet flaming among them. . . . I sat down on the rocky floor to enjoy the serene picture. Some crystal specks in the black ceiling high overhead, reflecting the light of a half-hid lamp, yielded this magnificent effect.
>
> I own, I did not like the cave so well for eking out its sublimities with this theatrical trick. But I have had many experiences like it, before and since; and we must be content to be pleased without too curiously analyzing the occasions.[9]

Jenny Lind

Swedish singer Jenny Lind was the world's most famous singer in the mid-1800s—so famous that a locomotive, clipper ships, songs, dances, soup, and a town were named after her. Even today, we call cribs with spiral posts "Jenny Lind cribs." In spite of all this fame, few people heard her voice because there was no way to record it back then.

strong Custer as a member of the grand duke's party. To Americans now, Custer is more famous than Grand Duke Alexis, but in 1872 visitors may have seen Custer as just another military man.

Custer was stationed in Elizabethtown, Kentucky, at the time he made his visit to the cave. He and his wife lived at the Hill House, now the Brown-Pusey Community House.[13]

A Union army general during the Civil War, Custer became the lieutenant-colonel of the Seventh Cavalry after the war. In 1876, he led an attack on Lakota Sioux, Cheyenne, and Arapaho warriors at Little Bighorn in Montana. Custer underestimated the American Indians' fighting abilities—they killed all 210 of Custer's men. His death, known as "Custer's Last Stand," made him famous.

Emperor Dom Pedro

Emperor Dom Pedro of Brazil visited Mammoth Cave around the 1880s. Accounts give more details about the emperor's stop for a drink of water on a stage ride from Cave City than about his trip in the cave. One article states that an officer told the locals, "Hats off while the sovereign drinks," so they took off their hats. When the officer offered the locals a drink, a Kentuckian said to the emperor, "Hats off while the sovereign drinks, we are all sovereigns in this country."[14]

Another version of the story says a man named S. J. Preston wouldn't remove his hat and told the emperor, "I am a Kentuckian and somewhat of a sovereign myself." The emperor was supposedly so impressed by Preston's pluck that he offered him a sugar plantation if he returned to Brazil with him.[15] Preston didn't take him up on the offer.

John Muir

Nineteenth-century conservationist, writer, and Sierra Club founder John Muir stopped at Mammoth Cave in 1867 as he walked from Indianapolis to Cedar Key, Florida, a trip he recorded in his posthumously published book *A Thousand-Mile Walk to the Gulf.* Muir wrote that a local school official suggested Muir stay near Mammoth Cave until the following spring, "assuring me that I would find much to interest me in and about the great cave," but he continued his walk to the gulf.[16]

Edwin Booth

Shakespearean actor Edwin Booth came from a theatrical family; his brother, John Wilkes, and their dad, Junius, were also actors. All three were famous in their day, but unless you're a buff of nineteenth-century theater, you know only John Wilkes Booth, who killed Abraham Lincoln in Ford's Theatre in 1865. John Wilkes's drop from actor to assassin set Edwin's acting career back temporarily, but Edwin eventually returned to the stage.

Edwin visited Mammoth Cave in 1876 and wrote to his daughter about his experience.

> Our guide was a bright young colored chap, who produced by his imitations of dogs, cows, etc., some fine effects of ventriloquism on our way through the cave. In pointing out to us a huge stone shaped like a coffin he would remark: "Dis is de giant's coff-in," then, taking us to the other dilapidated side of it: "Dis is what he coughed out." . . . [W]e went laughing at his weak jokes; for it was funny to us actors to this fellow throwing his wit at us, and our appreciation of his acting made him very happy.[17]

For years, cave guides could push a button in the cave chamber Booth's Amphitheatre to play a recording of Hamlet's "To be or not to be" soliloquy in remembrance of Edwin Booth. Today, the recording is a retired guide's doorbell! If Booth performed Hamlet or anything else in the cave, he didn't mention it in his letter.

Helen Miller Gould

Cave guide John Nelson took Helen Gould, philanthropist and daughter of railroad tycoon Jay Gould, into the cave in 1900. He asked Helen about her brother, probably George Jay Gould, who was also a railroad tycoon, though Nelson wasn't specific. When Helen asked Nelson if he knew who she was, he replied, "I think you're Miss Helen Gould, and if I'm not right, I have another thing coming." One of her six companions said, "Miss Helen, he's got us in here [the cave] and got us locked up, so we might as well acknowledge [it]." She was surprised that Nelson knew who she was because she hadn't registered at the hotel yet, so the desk clerk couldn't have told him.

A *Courier-Journal* reporter on the tour wrote an article stating that Nelson had proposed to Gould at the cave entrance!

Nelson and the cave made a good impression on Gould; she borrowed a Kodak camera and took his picture, told him she enjoyed the trip more than any trip she had ever made, and later sent him some books and a letter.[18]

Paul Von Hindenburg

John Nelson also took Prussian army officer Paul von Hindenburg into the cave. At the time of his visit in the early 1900s, Von Hindenburg was famous enough that Nelson knew who he was. Von Hindenburg later became more famous: first as commander of German forces in World War I and then as president of Germany from 1925 to 1934. Von Hindenburg's political party opposed the Nazis, but he also appointed Hitler chancellor in 1933.

Today, Americans know the name Von Hindenburg mostly from the zeppelin named after him. It became famous when it burst into flames while landing in New Jersey in 1937.

Von Hindenburg took a photograph of Nelson in the cave and sent it to Nelson from Germany with a letter of thanks.

Von Hindenburg and his companions built a monument in the cave, but soldiers tore it down after World War I.[19]

21

Colossal Cave Adventure

Cave Meets Computer

In 2007, I had a conversation with computer game enthusiast Jason Scott Sadofsky, who came to Mammoth Cave to make a documentary.

JASON: Hi, I'm Jason, digital historian. I'm going to Bedquilt Cave tomorrow to make a documentary on the early computer game *Colossal Cave Adventure* based on caving here at Mammoth Cave.
COLLEEN: Great, do you have any gear?
JASON: Ah, no.
COLLEEN: No problem, I'll lend you my spare helmet and light.
JASON: I'm a little worried. The cavers leading me in are all old guys; I don't know if they're up to such a strenuous trip.
COLLEEN: My husband, Rick Olson, is one of the cavers going with you.
JASON: He's young, right?
COLLEEN: He's fifty-seven.
JASON: Oh.

Jason really looked worried now.

The thirty-seven-year-old Jason and fifty-seven-year-old Rick had their own conversation at the Cave Research Foundation (CRF) headquarters that evening.

JASON: I've been working out at the gym to buff up for this cave trip. I can almost do a pull up.

RICK: OK, let us cavers carry your gear so you won't have to carry a pack. We're used to doing this.

Later, in Bedquilt Cave . . .

JASON: I'm exhausted; it must be time to head out of the cave.
RICK: Let's go a little farther. The ax head is just ahead; it's an important part of the game. You should film Dave [one of the cavers] with it; he bought his first computer just to play *Colossal Cave Adventure.*
JASON: You're right, it's an important part of the game. I can make it.

Later, at the ax head . . .

JASON: Great, I'm glad we made it to the ax head, but I'm beat and worried about being able to get out.
RICK: We're almost to the pit. The pit is an important part of the game; you have to see it.
JASON: You're right, it's an important part of the game. I can make it.

Later, at the pit . . .

RICK: If you brace yourself on this rock, you can lean over the edge and see fifty feet to the bottom.
JASON: No thanks, I can see it fine from over here.

Jason made it out of the cave, exhausted but alive, with footage to use for his documentary and an understanding of why real cavers are in better shape than gamers.

The First Computer Adventure Game

Much to the pleasure of cavers and computer geeks, the first computer adventure game, *Colossal Cave Adventure* (also known as *Adventure*), is based on cave exploration at Mammoth Cave. Caver and computer programmer Will Crowther created the game in 1975 based on his experience exploring the cave.

Gamers call this type of game a "text adventure," "adventure game," or "interactive fiction" (IF). IF is like reading a story, but you make decisions for the main character, which influences the game's direction and outcome.

The original *Adventure* was completely text based, but over the years gamers have added a few graphics and made other tweaks and changes.

Adventure and other IF games still have a following on the Internet, but the genre lost its commercial marketability when graphic video games became readily available in the mid-1980s.

Developed in 1976, *Adventure* was probably the first computer adventure or IF game, but other computer games had come before it. Computerized Tic-Tac-Toe, *Tennis for Two, Space War!,* and *Chase* were created in 1952, 1958, 1962, and 1967, respectively. Only the few computer students and professional computer programmers who existed then were exposed to these games. *Pong,* a video version of ping pong that even nongeeks remember, was created for arcades in 1972 and for home use in 1975.[1]

The Real Cave

Colossal Cave Adventure is based on exploration in Bedquilt Cave, a section of the Mammoth Cave system. The name "Bedquilt Cave Adventure" lacks the desired oomph for a computer game, though, which is probably why Will Crowther named the game after the adjoining Colossal Cave section.

Discovered around 1871, Bedquilt supposedly owes its name to the discovery of an Indian blanket near the entrance. One story goes that moonshiners hid their stills in Bedquilt.[2]

Many features in the game exist in the real cave. An ax head and a rusty iron rod possibly left by moonshiners lie in passages. Both cavers and gamers pass through Bird Chamber, the Hall of Mists, the Hall of the Mountain King, and a rock marked Y2. Gamers ponder what seems to them the mysterious meaning of the mark "Y2," but cavers know it's a survey notation for the second point on the Y survey.

Former CRF president Mel Park told a story about a computer gamer named Bev Schwartz, who joined CRF. Excited to be on an expedition to Bedquilt, she knew the names of many of the rooms and where passages went based on her experience playing *Adventure,* though she'd never been in the cave. Mel let Bev lead the caving party out; she found the way. Mel said, "The cave is a real maze, and this was an impressive accomplishment for a first-time visitor." Bev knew the route so well that Mel soon let her lead survey parties into Bedquilt.[3]

Cave Meets Computer

Will Crowther, a highly skilled computer programmer, helped create ARPANET (Advanced Research Projects Agency Network), the U.S. government's forerunner to the Internet in the 1960s. You would think that would be Will's claim to fame, but he said, "I've done all sorts of wonderful things in my career, [so] it's funny that the one thing I'm remembered for is *Adventure.*"[4]

Will and his wife, Pat, caved with CRF at Mammoth Cave in the early 1970s. Pat Crowther helped make history in 1972 by helping discover the connection between Flint and Mammoth Caves, making the whole system the longest in the world. As it turns out, though, Pat fell in love with trip leader John Wilcox, divorced Will, and married John.

Caving with his former wife and her new husband would have been awkward, so Will gave up caving. But thoughts of the cave stayed with him.

I asked Will what inspired him to create *Adventure.* He wrote to me,

> I was playing *Dungeons and Dragons* in an old, free-wheeling style with some office friends. [Dungeons and Dragons *is a role-playing fantasy game. Though not originally a computer game, it's now available as one.*] I was being paid to play with computers at my job, and I had time since I was going through a divorce. I believed I could make a computer do anything I wanted in those days. Also, I was in the process of entering years of accumulated Mammoth survey data into the computer, so I had the cave on my mind when I thought of computers.[5]

Around 1977, Stanford graduate student Don Woods got ahold of *Adventure* and expanded it. Not a caver, Don added some objects not found in the real Bedquilt Cave (or any other cave), including a dragon, a troll, a magic beanstalk, and a battery-dispensing vending machine.[6] Will's original version included magic words, ax-throwing dwarves, and a crystal bridge that magically appears, so some fantasy was already mixed in with his mostly realistic description of the cave.

Will told me he felt Don's changes to the game "were great fun. The game got more and more fanciful as you explored deeper, which was all to the good. He knew nothing about real caves of course. And he helped people port the game to many different computers, which was why it became a success."[7] The two met after the game became popular.

Playing *Adventure*

In *Adventure,* you get points by getting treasures and returning them to a room called the Well House. Along with run-of-the-mill treasures such as gold, silver, diamonds, emeralds, pearls, jewelry, and coins, you can acquire exotic things such as a Persian rug, a Ming vase (you have to pick up a pillow to cushion the fragile vase), spices, golden eggs, a pyramid, and a jewel-encrusted trident. Taking the pyramid gets you thirty-two points, but taking the diamonds is worth only five. Players also get one point for bringing issues of *Spelunker Today* magazine to a room called Witt's End (the room really exists, the magazine doesn't).

Dangers include ax-wielding dwarfs who want to kill you, a dragon, a pirate, and a snake that you can scare off with a little bird.

Adventure's Impact

When I asked Will about people's reaction to *Adventure*, he told me, "Most noncomputer people found it boring. Most computer people found it fascinating. Cavers liked it, but they seldom had access to it (computers then [in the 1970s] were as big as refrigerators and cost more than cars)."[8]

Will's daughters, Sandy and Laura, were about eight and five years old when their dad invented the game and so played it as kids. An article in *Digital Humanities Quarterly* said Sandy was addicted to the game, so I asked Will to ask her about this addiction. She told him,

> Yeah, I suppose you could say "addicted." To pay attention to *Adventure* long enough to solve it requires a fair degree of concentrated attention, and there was a period of time when I was having fun doing that.
>
> Laura and I worked together sometimes, with one of us mapping and the other typing. I had a great time with it then. I was still pretty young, I think—it's funny to remember adrenaline pumping like mad when a dwarf started throwing axes at me and the frustration with the twisty mazes. It felt very real.[9]

Soon after Will's family and friends engaged *Adventure*, college computer majors got ahold of it. Along with Don Woods and his fellow computer students at Stanford, students at MIT became obsessed with the game. Their fixation with solving the game supposedly slowed down MIT's computer department for a while.

Rick Olson, who helped guide documentary filmmaker Jason into Bedquilt Cave, remembers leaving the University of Illinois electron microscopy lab where he worked in the early 1980s to go to a CRF caving expedition at Mammoth Cave. When he saw how some students stayed late in the lab to use the computer (computers were few and far between back then) to play *Adventure,* he thought, "These poor guys are going to pretend to explore Mammoth Cave while I do it for real." Rick didn't play the game but knew about it because he had caved with Will Crowther in the 1970s.[10]

Words and phrases from *Adventure* have entered gamer vocabulary. The magic word *XYZZY,* used as a shortcut to transport the player between the spring house and the sand room, is especially popular; spellcheck even recognizes it! Other popular computer games, such as *Minesweeper* and *Zork* use it. Adventure fans at Goddard Space Flight Center used *XYZZY* as a command for a satellite.

Mathematician Ron Hunsinger claims *XYZZY* is a trick for remembering the following formula: "The cross product of two three-dimensional vectors is the vector whose length is the area of the parallelogram with the two given vectors as adjacent sides and direction perpendicular to the plane of that parallelogram."[11] I have no idea what that means, but I see why you need a memory trick for it. Crowther says he just made up the word *XYZZY* for *Colossal Cave Adventure.*

The magic word *plugh* transports players between Y2 and a place called Inside Building. Other games, including *Prisoner2, Haunted House,* and *Bedlam* adopted it.

Interactive fiction fans also like the phrase "Fee Fie Foe Foo." A deviation from the famous line spoken by the giant in "Jack in the Beanstalk," *Adventure* uses it as a magic phrase to transport golden eggs.

The sentence "You are in a maze of twisty little passages, all alike," referring to a confusing labyrinth in the game, became a catchphrase for computer geeks to refer to an argument that no one wins or a problem you can't solve.

Get Lamp

Jason borrowed my gear and went into Bedquilt with CRF cavers Rick Olson, Dave West, and Peter Bosted (the "old" cavers Jason was concerned about) as well as Mammoth Cave guide and caver Bruce Hatcher to film footage for *Get Lamp,* Jason's documentary on computer adventure games.[12] The DVD includes a feature on the real Bedquilt Cave as well as interviews with *Adventure* co-creator Don Wood, cavers Dave West, Roger Brucker, and Tommy Brucker, and several IF enthusiasts. The gamers talk about how *Adventure* gave birth to IF, IF's popularity in the 1970s and early 1980s, and its demise with the rise of graphic video games.

The title *Get Lamp* and the lantern on the DVD cover pay tribute to *Colossal Cave Adventure.* The lantern can be seen somewhere in the background during all the interviews.

Cavers and I laughed over Jason's worries about the old, decrepit cavers who wound up carrying his gear and coaxing him through Bedquilt Cave. The lesson: virtual caving and puzzle solving are fun and challenging, but real caving is far more challenging and will keep you in better shape—so play the game but also keep caving!

22

Cave Jokes

The Lowest Form of Humor

Mammoth Cave insurance plan: If you are killed or injured by a rock fall in the cave, you will be completely covered.

Mammoth Cave burial plan: If the cave collapses, you can't be buried any deeper any cheaper.

VISITOR: How often do rocks fall in the cave?
GUIDE: Once.

Some people call formations that go from the floor to the ceiling "pillars," but here in Kentucky we call them "columns" because we sleep on our pillars.

People used to think the saltpeter workers were miners, but really they were adults.

A young woman promised her mother she would never marry any man on the face of the earth. When she met the man of her dreams, she kept her promise to her mother by marrying the man in Mammoth Cave, under the earth. (*This joke goes back at least as far as 1870.*)

VISITOR: How come you don't have weddings in the cave anymore?
GUIDE: The park doesn't want to run marriage into the ground.

Hardy, har har! You crack me up!

RANGER FRANK HENRY: Ranger Keven lights the lanterns because he has a degree in lanternology. What was your lanternology professor's name?

RANGER KEVEN: Dr. Burns.
FRANK HENRY: And her first and middle names?
KEVEN: Kara Sene, Dr. Kara Sene Burns.

Some brewers from a big brewery came to Mammoth Cave and asked to buy some cave crickets. The cave wasn't a national park yet, so they were able to buy the crickets and take them with them. They kept coming back for more crickets. Finally, a guide asked them why they wanted cave crickets. The brewers said, "We need the crickets for their hops."

GUIDE: Here at the Water Clock, if you listen closely to the dripping water, you can tell what time it is. . . . Do you know what time it is?
VISITOR: Three o'clock?
GUIDE: No, it's time to go.

The ancient people who entered Mammoth Cave practiced a ritualistic game in the chamber behind Giant's Coffin. They stacked up rocks and tried to knock the stack down by rolling rocks at it. One person refused to participate in this ritual, so the chamber became known as the "Wouldn't Bowl Room" (a.k.a. the Wooden Bowl Room).

There has been much debate on whether the standing rock called "the Devil's Looking Glass" naturally fell that way or if the rock was stood up. We now know that it could not have been stood up because it has a date. (*The date 1812 is carved on the rock.*)

Two carrot people went on a cave tour. In spite of being carrots, they did fine until one of them hit his carrot top on a rock and knocked himself out. After carrying him out of the cave, the guides accompanied the injured carrot and his friend to the emergency room. After examining the carrot, the doctor told the guides and the other carrot person, "I have some good news and some bad news. Your carrot friend is going to live, but he's going to be a vegetable for the rest of his life."

John Houchen discovered Mammoth Cave when he shot a bear, wounded it, and chased it into the cave. We Kentuckians know Houchen was from Indiana (or Ohio or Michigan or some other state) because if he had been a Kentuckian, he would have killed the bear with the first shot.

Some people say John Houchen found the cave while hunting bear, but I think he had clothes on.

GUIDE AT STANDING ROCKS: If you look up at the rocks, you will see the witch on her broom. Can you see which way she is going?
VISITOR: She's going deeper into the cave.
GUIDE: Wrong, she's not going anywhere; she's stays right there.

VISITOR: Is it true that the military puts saltpeter in food to curb libido?
GUIDE: Well, I was in the navy twenty years, and it hasn't kicked in yet.

Acknowledgments

I owe a Mammoth-size thanks to all the people who made this book possible. Mammoth Cave ecologist Rick Olson provided many photos, helped me interpret scientific papers, researched limestone dating, proofread the manuscript, and, as my husband, provided moral support throughout the project. Rick Toomey, director of the Mammoth Cave International Center for Science and Learning, all-around problem solver, and friend, provided much scientific knowledge. Geologist and caver extraordinaire Art Palmer shared photos and scientific information. U.S. Geological Survey earthquake guru Lucy Jones was kind enough to answer my seismic questions. George Corrie, Lewis Cutliff, Joe Duvall, Reverend Fred Haney Jr., Coy Hanson, Elisa Kagan, Brice Leech, George McCombs, Keven Neff, Louie B. Nunn, Mike Wiles, and Rachel Wilson shared their stories with me.

Many more people inspired me in writing this book. If I forgot to mention you, please forgive me.

Notes

1. The Mammoth Cave Dating Guide

1. Darryl E. Granger and Derek Fabel, "Cosmogenic Isotope Dating," in *Encyclopedia of Caves,* ed. David Culver and William B. White (San Diego: Elsevier Academic Press, 2005), 138.

2. Mary C. Kennedy and Patty Jo Watson, "The Chronology of Early Agriculture and Intensive Mineral Mining in the Salts Cave and Mammoth Cave Region, Mammoth Cave National Park, Kentucky," *Journal of Cave and Karst Studies* 59, no. 1 (1997): 5.

3. "Basic Principles of Radiocarbon Dating," National Science Foundation Arizona AMS Facility, n.d., at http://www.physics.arizona.education/theory.htm (accessed May 2006).

4. Thomas Higham, "The 14C Method," n.d., at http://www.c14dating.com/int.html (accessed January 2006).

5. D. Fleitmann, S. J. Burns, A. A. Al-Subbary, M. A. Al-Aowah, J. Kramers, and A. Matter, "Uranium-Series Dating of Stalagmites from Socotra, Yemen," paper presented at the Yemen Science Conference, Taiz, Yemen, October 11–13, 2002.

6. Dr. Jeffrey Dorale, email message to the author, October 4, 2005.

7. John W. Hess and Russell S. Harmon, "Geochronology of Speleothems from the Flint Ridge Mammoth Cave System, Kentucky, U.S.A.," in *Proceedings of the Eighth International Congress of Speleology, Bowling Green, Kentucky, 1981* (Huntsville, Ala.: National Speleological Society, 1981), 435.

8. Fleitmann et al., "Uranium-Series Dating of Stalagmites from Socotra, Yemen."

9. Jeffrey A. Dorale, R. Lawrence Edwards, E. Calvin Alexander Jr., Chuan-Chou Shen, David A. Richards, and Hai Cheng, "Uranium-Series Dating of Speleothems: Current Techniques, Limits, and Applications," in *Studies of Cave Sediments: Physical and Chemical Records of Paeoclimate,* ed. J. E. Mylroie (New York: Kluwer Academic and Plenum, 2004), 177–82.

10. Granger and Fabel, "Cosmogenic Isotope Dating," 138–41.

11. Darryl E. Granger, Derek Fabel, and Arthur N. Palmer, "Plio-Pleistocene Incision of the Green River, Kentucky, from Radioactive Decay of Cosmogenic 26Al and 10Be in Mammoth Cave Sediments," *GSA Bulletin* 113 (2001): 825–36.

12. E. R. Pohl, "Upper Mississippian Deposits in South Central Kentucky," *Transactions of the Kentucky Academy of Sciences* 31 (1970): 15; H. Willman, E. Atherton, T. Buschbach, C. Collinson, J. Frye, M. Hopkins, J. Lineback, and J. Simo, "Handbook of Illinois Stratigraphy," *Illinois State Geological Survey Bulletin* 95 (1975): 140–58.

13. F. Devuyst, L. Hance, H. Hou, X. Wu, S. Tian, M. Coen, and G. Sevasatopulo, "A Proposed Global Stratotype Section and Point for the Base of the Visean Stage (Carboniferous), the Penchong Section, Guangxi, South China," *Episodes* 26, no. 2 (2003): 105–15.

14. C. Rexroad and R. Liebe, "Conodonts from the Paoli and Equivalent Formations in the Illinois Basin," *Micropaleontology* 8, no. 4 (1962): 509–14.

15. E. Trapp and B. Kaufmann, "Hochprazise U-Pb datierungen von Pyroklasticka im Jungpalaozoikum," *Neustadt; Schwerpunkprogramm der deutschen Forschungsgemeinschaft DFG* 1054 (2002): 18–19.

2. Rockphobia

1. Richard L. Powell, "A Report from the Consulting Geologist: The Rockfall of January 1994 in the Rotunda, Mammoth Cave, Kentucky," unpublished geological report, 1994, Mammoth Cave National Park Manuscript Collection, Mammoth Cave, Ky., 1.

2. Ibid., 6.

3. Rick Olson, "This Old Cave: The Ecological Restoration of the Historic Entrance of Mammoth Cave, and Mitigation of Visitor Impact," unpublished paper, 1996, Mammoth Cave National Park Manuscript Collection.

4. Dr. Arthur Palmer, State University of New York, Oneonta, email message to the author, October 3, 2002.

5. Dr. Lucile M. Jones, science adviser for Risk Reduction, U.S. Geological Survey, email message to the author, May 11, 2012.

6. Brice Leech, Mammoth Cave National Park natural-resources specialist, email message to the author, May 22, 2012.

7. Mike Wiles, Jewel Cave National Monument cave-management specialist, email message to the author, October 5, 2006.

8. Nahum Ward, "Plan and Description of the Great and Won-

derful Cave in Warren County, Kentucky," *Scots Magazine & Edinburgh Literary Miscellany,* December 1816, 909–14.

9. Elisa Kagan, Institute of Earth Sciences, Hebrew University of Jerusalem, email message to the author, May 8, 2007.

10. U.S. Geological Survey, "Historic Earthquakes," n.d., at http://earthquake.usgs.gov/earthquakes/states/events/1811-1812.php/ (accessed September 2005).

11. Jake Page and Charles Officer, *The Big One: The Earthquake That Rocked Early America and Helped Create a Science* (New York: Houghton Mifflin, 2004), 16.

12. U.S. Geological Survey, "Largest Earthquakes in the United States," n.d., at http://earthquake.usgs.gov/earthquake/states/10_largest_us.php/ (accessed September 2005).

13. U.S. Geological Survey, "Historic Earthquakes."

14. Page and Officer, *The Big One,* 16; Ronald L. Street and Otto W. Nuttli, "The Great Central Mississippi Valley Earthquakes of 1811–1812," *Kentucky Geological Survey,* Special Publication 14, series 11 (1990): 3.

15. Page and Officer, *The Big One,* 7; Street and Nuttli, "The Great Central Mississippi Valley Earthquakes of 1811–1812," 1.

16. Page and Officer, *The Big One,* 203.

17. Arch C. Johnston and Eugene S. Schweig, "The Enigma of the New Madrid Earthquakes of 1811–1812," *Annual Review of Earth Planet Science* 24 (May 1996): 339–84.

18. Kaye M. Shedlock, "Intraplate Earthquakes," n.d., AccessScience@McGraw-Hill, at http://www.accessscience.com (accessed September 2005).

19. D. Kolata and T. Hildebrand, "Structural Underpinnings and Neotectonics of the Southern Illinois Basin: An Overview," *Seismological Research Letters* 68 (1997): 499–510.

20. Rick Olson, Mammoth Cave National Park ecologist, and Dr. Rick Toomey, Mammoth Cave National Park paleontologist, email message to the author, November 2012.

21. Sammuel V. Panno, C. C. Lundstrom, Keith C. Hackley, B. Brandon Curry, B. W. Fouke, and Z. Zhang, "Major Earthquakes Recorded by Speleothems in Midwestern U.S. Caves," *Bulletin of the Seismological Society of America* 99 (2009): 2147.

3. Bones

1. Ronald Wilson, National Park Service paleontologist, email message to the author, May 1997.

2. Elizabeth Hill, collection manager, Vertebrate Paleontology Section, Carnegie Museum of Natural History, email message to the author, May 1997.

3. Rick Toomey, Illinois State Museum paleontologist, email message to the author, May 1997; Ronald Wilson, "Vertebrate Remains in Kentucky Caves," in *Caves and Karst in Kentucky,* ed. Percy H. Dougherty (Lexington: University Press of Kentucky, 1985), 172–73.

4. Mona L. Colburn, "Paleontological Inventory Project: Vertebrate Remains Found in Select Passages and Caves at Mammoth Cave National Park, Kentucky," unpublished report, 2005, Illinois State Museum, Springfield.

5. Ibid., 287.

6. Ibid., 288–90.

7. Quoted in Colburn, "Paleontological Inventory Project," 239.

8. Ibid., 239–42.

9. Ibid., 257–58.

10. Ibid., 282–83.

11. Ibid., 256–57.

12. William H. Kern, "Southeastern Pocket Gopher," University of Florida Institute of Food and Agricultural Sciences Extension, 1991, at http://www.edis.ifas.ufl.edu/pdffiles/uw/uw08100.pdf.

13. Colburn, "Paleontological Inventory Project."

14. Ibid., 260–61, quoting Bjorn Kuton and Elaine Anderson, *Pleistocene Mammals of North America* (New York: Columbia University Press, 1980).

15. Ibid., 264.

16. Ibid., 225.

17. Ibid., 279–80.

18. Ibid., 293.

19. Ibid.

20. Ibid., 121–24.

4. Secret Lives of Cave Critters Revealed!

1. Thomas L. Poulson, "Cave Animals of Mammoth Cave National Park," unpublished paper, 1993, Mammoth Cave National Park Manuscript Collection, Mammoth Cave, Ky., 2.

2. Monisha M. Louis, "Age, Growth, and Fin Erosion of the Northern Cavefish, Amblyopsis Spelaea, in Kentucky and Indiana," master's thesis, University of Louisville, 1999.

3. Rick Olson, Mammoth Cave National Park ecologist, and Dr.

Horton H. Hobbs, Wittenberg University biologist, personal communication to the author, July 2010.

4. Dr. Thomas Poulson, University of Illinois biologist, personal communication to the author, July 2010.

5. Luis Espinasa and Monika Espinasa, "Why Do Cave Fish Lose Their Eyes?" *Natural History* 6 (2005): 48–49.

6. "Fish Lateral Line System," n.d., at http://lookd.com/fish/laterallinesystem.html (accessed September 2009).

7. David C. Culver and Tanja Pipen, *The Biology of Caves and Subterranean Habitats* (Oxford: Oxford University Press, 2009), 114.

8. Poulson, "Cave Animals of Mammoth Cave National Park," 26.

9. Ibid.

10. Anne Marie Helmenstine, "If I Leave My Goldfish in the Dark, Will It Turn White?" n.d., at http://chemistry.about.com/od/chemistryfaqs/f/goldfish.htm (accessed September 2009).

11. Poulson, "Cave Animals of Mammoth Cave National Park," 30.

12. Horton H. Hobbs, Wittenberg University biologist, email message to the author, April 10, 2010.

13. Lon A. Wilkens and James L. Larimer, "Photosensitivity in the Sixth Abdominal Ganglion of Decapod Crustaceans: A Comparative Study," *Journal of Comparative Physiology* 106 (1976): 69.

14. Thomas L. Poulson, "Animals in Aquatic Environments: Animals in Caves," in *Handbook of Physiology,* ed. D. B. Dill (Oxford: Oxford University Press, 1988), 755.

15. Ibid.

16. Poulson, "Cave Animals of Mammoth Cave National Park," 9.

17. Arthur T. Leitheuser, "Ecological Analysis of the Kentucky Cave Shrimp, *Palaemonias ganteri hay,* Mammoth Cave National Park," *Central Kentucky Cave Survey Bulletin* 1 (1984): 73.

18. Ibid., 75.

19. Rick Olson, Mammoth Cave National Park ecologist, personal communication with the author, August 2010.

20. Poulson, "Cave Animals of Mammoth Cave National Park," 8.

21. Ibid., 6.

22. Michael Sutton, "Cave Fauna of the Mammoth Cave Region," unpublished paper, 1995, Mammoth Cave National Park Manuscript Collection.

23. J. Fawley, "*Eurycea lucifuga,*" Animal Diversity Web, 2002, at http://animaldiversity.ummz.umich.edu/site/accounts/information/Eurycea_lucifuga.html.

24. Culver and Pipan, *The Biology of Caves and Other Subterranean Habitats*, 95.

25. Kathleen H. Lavoie, Kurt L. Helf, and Thomas L. Poulson, "The Biology and Ecology of North American Cave Crickets," *Journal of Cave and Karst Studies* 69, no. 1 (2007): 123.

26. Ibid., 121.

27. Ibid., 123.

28. "Cave Crickets Monitoring," National Park Service Inventory & Monitoring, 2008, at http://science.nature.gov; Poulson, "Cave Animals of Mammoth Cave National Park," 19.

29. Dr. Thomas Poulson and Dr. Kurt Helf, personal communication with the author, May 2009.

30. Lavoie, Helf, and Poulson, "The Biology and Ecology of North American Cave Crickets," 116.

31. Ibid., 120–21.

32. Eugene H. Studier and Kathleen H. Lavoie, "Leg Attenuation and Seasonal Femur Length: Mass Relationships in Cavernicolous Crickets (Orthoptera: Gryllidae and Rhaphidophoridae)," *Journal of Cave and Karst Studies* (August 2002): 127.

33. T. C. Kane and T. L. Poulson, "Foraging by Cave Beetles: Spatial and Temporal Heterogeneity of Prey," *Ecology* 57 (1976): 799.

34. Dr. Kurt Helf, personal communication with the author, August 2009.

35. Culver and Pipan, *The Biology of Caves and Other Subterranean Habitats*, 95–96.

36. Sutton, "Cave Fauna of the Mammoth Cave Region"; Poulson, "Cave Animals of Mammoth Cave National Park," 24.

37. Poulson, "Cave Animals of Mammoth Cave National Park," 21.

38. Sutton, "Cave Fauna of the Mammoth Cave Region."

39. Ibid.; Poulson, "Cave Animals of Mammoth Cave National Park," 18.

40. Sutton, "Cave Fauna of the Mammoth Cave Region."

41. Ibid.

42. "Allegheny Woodrat," Inside Natural Resources, Intranet Portal for Natural Resources, n.d., at http://www1.nps.gov/im/units/CUPN/monitor/woodrats/woodrats.cfm (accessed September 2009).

43. Ibid.

44. Anne-Marie Monty and George A. Feldhamer, "Conservation Assessment for the Eastern Woodrat and the Allegheny Woodrat," U.S. Department of Agriculture, Forest Service, Eastern Region, May 2002, 19.

45. Ibid.

46. Steve Thomas, Cumberland Piedmont Network, email message to the author, October 21, 2010.

47. Ibid.

48. Ibid.

49. Ibid.

50. Rick Olson, "This Old Cave: The Ecological Restoration of the Historic Entrance Ecotone of Mammoth Cave and Mitigation of Visitor Impact," unpublished paper, 1996, Mammoth Cave National Park Manuscript Collection.

51. John E. Hill and James D. Smith, *Bats: A Natural History* (Austin: University of Texas Press, 1982), 108, 185, 191.

52. Anne-Marie Hodge, "Social Clocks: How Do Cave Bats Know When It Is Dark Outside?" *SciLogs,* August 17, 2009, at http://scienceblogs.com/clock/2008/04/17/how-do-bats-in-a-cave-know-if/.

53. Centers for Disease Control and Prevention, "Rabies in the U.S. and around the World," at http://www.cdc.gov/rabies/location/index.html (accessed August 2016).

54. Ibid.

55. Ibid.

56. Ibid.

57. Centers for Disease Control, "Histoplasmosis," n.d., at http://www.cdc.gov/fungal/diseases/histoplasmosis/index.html (accessed October 2009).

58. National Eye Institute, "Facts about Histoplasmosis," n.d., at https://nei.nih.gov/health/histoplasmosis/histoplasmosis (accessed August 2016).

59. Dr. Rickard Toomey, Mammoth Cave National Park paleontologist, personal communication with the author, July 25, 2009.

60. Dr. Rickard Toomey, Mammoth Cave National Park paleontologist, personal communication with the author, September 9, 2015.

61. Turkey Vulture Society, homepage, n.d., http://www.fcps.edu/islandcreekes/ecology/turkey_vulture.htm (accessed August 2016).

62. Chipper Woods Bird Observatory, "Eastern Phoebe," n.d., at http://www.wbu.comchipperwoods/photos/ephoebe.htm (accessed October 2009).

63. "Bear Essentials of Hibernation," n.d., at http://www.pbs.org/wgbh/nova/nature/bear-essentials-of (accessed August 2016).

64. Kentucky Department of Fish and Wildlife Resources, "Black

Bears in Kentucky," n.d., at Fw.ky.gov/Wildlife/Pages/Black-Bears .aspx (accessed October 2009).

5. The Gypsum–Twinkie Connection

1. "History of Engineering: Building the Pyramids," n.d., at http://academic.pgcc.edu (accessed April 2003).

2. "Why Is Plaster of Paris Called So?" *Times of India,* December 16, 2007, at http://timesofindia.indiatimes.com/home/sunday-times/Why-is-the-plaster-of-Paris-called-so/articleshow/2624945.cms.

3. Department of Geography, Michigan State University, "Gypsum," n.d., at http://www.geo.msu.edu/geo333/gypsummining.html (accessed April 2003).

4. Tuan Liu, quality-control manager, Redhook Ale Brewery, Woodinville, Wash., email message to the author, March 29, 2003.

5. Interstate Brand Corporation (makers of Hostess products), email message to the author, April 3, 2003.

6. Procter & Gamble customer service representative, email message to the author, April 18, 2003; Colgate-Palmolive customer service representative, email message to the author, April 20, 2003.

7. Steve Jacobs, Procter & Gamble Company, via Howard Kalnitz, email message to the author, June 5, 2003.

8. "What Is Tooth Powder?" Roxana Lo. Dental Blog, n.d., at https://santarosadentist.wordpress.com/2013/05/21/what-is-tooth-powder/ (accessed August 2016).

9. Tom's of Maine, "Product Details," n.d., http://www.tomsofmaine.com/product-details/anticavity-fluoride-toothpaste (accessed August 2016).

10. FGD Products, "Introduction to FGD Gypsum," n.d., at http://www.fgdproducts.org/FGDGypsumIntro.htm (accessed August 2016).

11. Western Minerals, "About Western Minerals," n.d., http://www.westernminerals.com/aboutwm.php?PHPSESSID=5049e40f03367c4c1abb16d633447217 (accessed August 2016).

12. Jack Kelly, *Gunpowder* (Cambridge, Mass.: Basic Books, 2004), 2.

13. Phil Earnshaw, "A History of Gunpowder," Sharpshooter, n.d., at http://www.sportshooter.com/reloading/historygunpowder.htm (accessed September 2003).

14. Epsom Salt Industry Council, home page, n.d., http://www.epsomsaltcouncil.org (accessed September 2003).

15. "Preparations for the Lewis and Clark Expedition," Monticello.org, n.d., at https://www.monticello.org/site/research-and-

collections/preparations-lewis-and-clark-expedition (accessed August 2016).

6. Prairie Park Companion

1. G. R. Wilkins, P. A. Delcourt, H. R. Delcourt, F. W. Harrison, and M. R. Turner, "Paleoecology of Central Kentucky since the Last Glacial Maximum," *Quaternary Research* 36 (1991): 224–39.

2. Rick Olson, "Torch Fuels Used by Prehistoric Indian Cavers: Their Utility and Botanical Significance," in *Proceedings of Mammoth Cave National Park's Seventh Annual Science Conference* (Mammoth Cave, Ky.: Mammoth Cave National Park, 1998), 5–8.

3. P. J. Watson, *Archaeology of the Mammoth Cave Area* (New York: Academic Press, 1974), 170.

4. Randy Seymour, *Wildflowers of Mammoth Cave National Park* (Lexington: University Press of Kentucky, 1997), 117.

5. Rick Olson, "The Ecological Foundation for Prescribed Fire in the Mammoth Cave Area," *George Wright Forum* 22, no. 3 (2005): 25.

6. G. Williams, "References on the American Indian Use of Fire in the Ecosystem," U.S. Forest Service, 1994, at http://www.wildlandfire.com/docs/2003_n_before/Williams_Bibliography_Indian_Use_of_Fire.pdf.

7. Quoted in Reuben Gold Thwaites, *Early Western Travels 1748–1846* (Cleveland: Clark, 1904), 92.

8. Quoted in ibid., 92.

9. Alexander Bullitt, *Rambles in the Mammoth Cave* (1845; reprint, New York: Johnson Reprint Corporation, 1973), vi.

10. John Hussey, "Report on the Botany of Barren and Edmonson Counties," *Kentucky Geological Survey Part II* 1 (1876): 12.

11. Communication from Randy Seymour via Rick Olson, Mammoth Cave National Park ecologist, personal communication with the author, August 2016; see also Seymour, *Wildflowers of Mammoth Cave National Park*.

12. Robert Davidson, *An Excursion to the Mammoth Cave and the Barrens of Kentucky with Some Notices of the Early Settlement of the State* (Lexington, Ky.: Skillman & Son, 1840), 29–30.

13. Ibid., 30.

14. Charles E. Kay, "Aboriginal Overkill and Native Burning: Implications for Modern Ecosystem Management," *Western Journal of Applied Forestry* 104 (1995): 126.

15. Randy Seymour, "A Contribution to the Recorded Flora of

Mammoth Cave National Park," paper presented at the Fifth Annual Mammoth Cave National Park Science Conference, 1996, Mammoth Cave, Ky., 2, copies of the proceedings donated by Copier Imaging Systems, Inc., Bowling Green, Ky.

7. Prehistoric Cavers

1. Patty Jo Watson, "Prehistoric Cultural Debris from the Vestibule Trenches," in *Archeology of the Mammoth Cave Area,* ed. Patty Jo Watson (New York: Academic Press, 1974), 95.

2. Philip DiBlasi, archaeologist, University of Louisville, personal communication with the author, 2006.

3. Patty Jo Watson, *The Prehistory of Salts Cave Kentucky* (Springfield: State of Illinois, 1969), 60.

4. Frederick Benington, Carl Melton, and Patty Jo Watson, "Carbon Dating Prehistoric Soot from Salts Cave, Kentucky," *American Antiquity* 28 (1962): 241.

5. Kenneth B. Tankersley, "Prehistoric Mining in the Mammoth Cave System," in *Of Caves & Shell Mounds,* ed. Kenneth C. Carstens and Patty Jo Watson (Tuscaloosa: University of Alabama Press, 1996), 37.

6. Watson, *The Prehistory of Salts Cave Kentucky,* 58.

7. George Crothers, archaeologist, University of Kentucky, personal communication with the author, May 2001.

8. Mary C. Kennedy and Patty Jo Watson, "The Chronology of Early Agriculture and Intensive Mineral Mining in the Salts Cave and Mammoth Cave Region, Mammoth Cave National Park, Kentucky," *Journal of Cave and Karst Studies* 59 (1997): 5.

9. Thomas Higham, "The 14C Method," c14dating, n.d., at http://www.c14dating.com/int.html (accessed April 2001).

10. Dr. Patty Jo Watson, archaeologist, personal communication with the author, April 2003.

11. Guy Prentice, "Archeological Overview and Assessment of Mammoth Cave National Park," *National Park Service Archeological Service* 1 (1993): 18–19.

8. Saltpeter

1. Anonymous sailor, personal communication with the author, 2010.

2. Dr. Stan Sides, email message to the author, October 19, 2003.

3. Matthew Seelinger, historian, Army Historical Foundation, email message to the author, October 3, 2003.

4. Michael J. Crawford, head of the Early History Branch, Naval Historical Center, email message to the author, November 4, 2003.

5. M. G. Kafkalas, U.S. Army Heritage and Education Center, email message to the author, October 22, 2003.

6. Quoted in Paul Russell Cutright, "I Gave Him Barks and Saltpeter," *American Heritage* 15, no. 1 (1963), at http://www.americanheritage.com/content/%E2%80%9Ci-gave-him-barks-and-saltpeter-%E2%80%A2%E2%80%9D.

7. Caring Medical of Oak Park, Illinois, "Prolotherapy," *Prolonews*, n.d., at http://www.prolonews.com (accessed June 2003).

8. Jack Kelly, *Gunpowder* (New York: Basic Books, 2004), 36, 62–63.

9. Anne Marie Helmenstine, "Gunpowder Facts and History," n.d., at http://chemistry.about.com (accessed June 2003).

10. Kelly, *Gunpowder,* 17.

11. Bacon quoted in ibid., 25.

12. Lazarus Ercker, *Lazarus Ercker's Treatise on Ores and Assaying,* trans. Anneliese Grünhaldt Sisco and Cyril Stanley Smith (Chicago: University of Chicago Press, 1951), 292, 293.

13. Kelly, *Gunpowder,* 36–37.

14. Ercker, *Treatise on Ores and Assaying,* 306.

15. Kelly, *Gunpowder,* 5–6.

16. Ibid., 160.

17. Peter M. Hauer, "Saltpetre History," *Journal of Spelean History* 16, no. 2 (1975): 17.

18. Kelly, *Gunpowder,* 171–73.

19. Document recording a survey on September 3, 1799, for Valentine Simmons, June 3, 1800, entry in Surveyor's Book A, pp. 268–69, Warren County Courthouse, Bowling Green, Ky.

20. Harold Meloy and William R. Halliday, "A New Concept of the Initial History of Mammoth Cave, 1798–1812," *Journal of Spelean History* 1, no. 4 (1968): 113.

21. Deed Book 6, pp. 48, 49, 52, Warren County Clerk's Office, Bowling Green, Ky.

22. "Charles Wilkins," unpublished paper, Mammoth Cave National Park Manuscript Collection, Mammoth Cave, Ky.

23. "Fleming Gatewood," unpublished paper, Mammoth Cave National Park Manuscript Collection.

24. "Charles Wilkins," Mammoth Cave National Park Manuscript Collection.

25. Ibid.

26. Carol A. Hill and Duane DePaepe, "Saltpeter Mining in Kentucky Caves," *The Register* 77 (1979): 249–50.

27. Thomas Jefferson, *Notes on the State of Virginia* (Richmond, Va.: Randolph, 1853), 278.

28. Thomas Jefferson to Pierre S. du Pont de Nemours, February 12, 1806, Thomas Jefferson Papers Series, Library of Congress, Washington, D.C.

29. Unknown letter writer to Archibald McCall, January 7, 1813, Mammoth Cave National Park Manuscript Collection, emphasis added.

30. Archibald McCall to unknown recipient, Philadelphia, January 29, 1814, Mammoth Cave National Park Manuscript Collection.

31. Douglas Plemons, "The United States Saltpeter Cave Survey, 2006," *Journal of Spelean History* 41, no. 2 (2007): 47–66.

32. Quoted in Don Fig and Gary Knudsen, "Niter Mining: An Incipient Industry of the Red River Gorge, Kentucky," in *Proceedings of the Symposium on Ohio Valley Urban and Historic Archaeology* (Louisville, Ky.: Archaeological Survey, University of Louisville, 1984), 67–68.

33. Jefferson to du Pont de Nemours, February 12, 1806.

34. Dr. Thomas Mackey, University of Louisville, email message to the author, September 6, 2007.

35. Kelly King Howes, *War of 1812* (Detroit: Gale Group, 2002), 13–14, 26, 31–32.

36. Kelly, *Gunpowder*, 227–29.

37. Ibid., 131.

38. Ibid., 235.

9. The Cave Cure

1. Dr. John Croghan to General Jesup, January 13, 1843, Mammoth Cave National Park Manuscript Collection, Mammoth Cave, Ky.

2. Tami Port, "Chronic Diseases Caused by Mycobacteria: Tuberculosis and Leprosy Bacterial Infection," n.d., at http://human-infections.suite101.com (site no longer available; accessed September 2007).

3. American Lung Association, "Tuberculosis," November 2007, http://www.lung.org/lung-health-and-diseases/lung-disease-lookup/tuberculosis/ (accessed August 2016).

4. Ibid.

5. Stephen L. Tankersley, "Detection and Recovery of Anthro-

pogenic Introduced Pathogenic Microorganisms in Mammoth Cave, Mammoth Cave National Park, Kentucky," master's thesis, Wright State University, 1993, iii–iv.

6. Quoted in Dr. Stanley D. Sides, email message to the author, June 30, 2009.

7. Jill Nolt, "The Sanatorium Landscape," n.d., at http://www.faculty.virginia.edu/blueridgesanatorium/landscape.html (accessed November 2004).

8. Yoshifumi Miyazaki, "Science of Natural Therapy," n.d., at https://www.marlboroughforestry.org.nz/mfia/docs/naturaltherapy.pdf (accessed August 2016), 11.

9. Ebenezer Meriam, "Mammoth Cave," *New York Municipal Gazette,* February 21, 1844.

10. Robert Montgomery Bird, *Peter Pilgrim; or A Rambler's Recollections* (Philadelphia: Lea & Blanchard, 1838), 109.

11. Quoted in Scott E. Sallee, "Tambo," unpublished paper, Mammoth Cave National Park Manuscript Collection.

12. Horace C. Hovey, *Guide Book to the Mammoth Cave of Kentucky* (Cincinnati: Clark, 1887), 30.

13. Charles W. Wright, *A Guide Manual to the Mammoth Cave of Kentucky* (Louisville, Ky.: Bradley & Gilbert, 1860), 15.

14. Trevor R. Shaw, *History of Cave Science* (Sydney: Sydney Speleological Society, 1992), 223–24.

15. Patricia Lewis, owner and operator of Free Enterprise Radon Health Mine, email message to the author, February 2004.

16. Bobby Carson, Mammoth Cave National Park, email message to the author, February 2004.

17. Free Enterprise Health Mine, "Meet Our Clients," n.d., at http://www.radonmine.com/clients.php (accessed May 2004).

18. U.S. Environmental Protection Agency, "Radon," n.d., at http://www.epa.gov/radon (accessed May 2004).

19. Ryan L. Dansak, "Mining for Relief," *RT Image,* December 11, 2000.

20. Dr. Stan Sides, member of the Speleotherapy Congress and attendee of the Twelfth International Speleotherapy Symposium at Aggtelek National Park in Hungary, email message to the author, September 2001.

21. Dr. Stanley Sides, email message to the author, April 2004.

22. Tibor Horvath, "Speleotherapy: A Special Employment of the Cave Microclimate," n.d., copy provided to the author by Katalin Bol-

ner Takacsne, Speleological Department, Hungarian Ministry for Environment, 2004.

23. E. Weigl, Z. Hermanova, and J. Szotkowska, "Immunological Analysis of Children's Asthmatic Patients after Speleotherapy," in *Proceedings from the 11th International Symposium of Speleotherapy* (Zlaté Hory, Czech Republic: Eddel, 1999), 105.

24. Gasteiner Heilstollen, "Pricelist," n.d., at http://www.gasteiner-heilstollen.com/en/prices-packages/ (accessed March 2004).

25. Julia Hakobyan, "Speleotherapy: Salt of the Earth Helps Asthma Patients Breath Easy," *Armenia Week*, September 2001.

26. Sylvia Beamon, "Gone to the Salt Mines—Whatever For?" Souterrains, n.d., at http://wwwlxs4all.nl/~jorbons/souterrains/art/saltrome.html (accessed September 2001).

27. Birstono Sanatorija Versme, home page, n.d., at http://www.versme.com/go.php/lit/English/1 (accessed September 2001).

28. See the website for a company that sells salt lamps for home use at http://www.saltlamp.com (accessed September 2001).

29. Horvath, "Speleotherapy," 129–30.

10. Exploring the World's Longest Cave

1. Quoted in Bernd Kliebhan, "The Contribution of E. A. Martel (1859–1938) to the Development of Caving Technique," trans. Dr. Klaus Thomas, *Mill. Verb. Dt. Höhlen—u. Kartstforsch* 45, no. 2 (1999), at http://carpediem.kliebhan.de/spelhist/mar/eam-contrib.html.

11. The Cave Wars

1. "A Windy Controversy Ends on a Note of Harmony," *Louisville Courier-Journal*, December 2, 1940.

2. "Unscrupulous Misrepresentations Regarding Conditions at the Mammoth Cave," *Evansville Press*, circa 1923, photocopy of a newspaper clipping, Mammoth Cave National Park Manuscript Collection, Mammoth Cave, Ky.

3. Joe Duvall, interviewed June 1999, Cave City, Ky.; George McCombs, interviewed June 1999, Munfordville, Ky.; and Louie B. Nunn, interviewed August 1999, Horse Cave, Ky.

18. Nuclear-Fallout Shelters in Mammoth Cave

1. R. H. Hayes, chief of engineering, Division of U.S. Army Engineer District, to George Davis, director of Edmonson County Civil

7. Herman Melville, *Mardi and a Voyage Thither* (New York: Harper and Brothers, 1849), 267.

8. Herman Melville, *The Confidence-Man* (London: Longman, Brown, Green, Longmans & Roberts, 1857), 143.

9. Jules Verne, *Journey to the Center of the Earth* (1864; reprint, Pleasantville, N.Y.: Reader's Digest Association, 1992), 152.

10. Jules Verne, *All Around the Moon* (Paris: Hetzel, 1870), 270, at http://www.freeclassicebooks.com/Jules%20Verne/Jules%20 Verne%20-%20English/All%20Around%20the%20Moon.pdf.

11. Jules Verne, *The Mysterious Island* (1874; reprint, New York: Scribner's, 1920), 450.

12. Jules Verne, *The Underground City* (1877; reprint, Minneapolis: Filiquarian Publishing, 2006), 68–69, 100.

13. Jules Verne, *Facing the Flag* (Chicago: F. T. Neely, 1894), 109.

14. Ibid.

15. Jules Verne, *Will of an Eccentric* (Paris: Hetzel, 1900), 266.

16. John Uri Lloyd, *Etidorhpa* (Cincinnati: Robert Clarke, 1896), 91.

17. Joseph H. Cater, *The Ultimate Reality,* vol. 1 (Pomeroy, Wash.: Health Research Books, 1998), 95.

18. L. Frank Baum, "The Box of Robbers," in *American Fairy Tales,* unabridged (1901; reprint, Tarzana, Calif.: Lanval, n.d.), 11, at https:// books.google.com/books?id=ZanaweDCZfEC&pg=PA2&dq=Baum,+ American+Fairy+Tales&hl=en&sa=X&ved=0ahUKEwjX9OvRiNXO AhXDQyYKHaT8D7AQ6AEIOzAE#v=onepage&q=Baum%2C%20 American%20Fairy%20Tales&f=false.

19. Edward Stratemeyer, *An Undivided Union* (1899; reprint, Whitefish, Mont.: Kessinger, 2015), 246.

20. Irving Bacheller and William Harlowe, *Keeping Up with Lizzie* (New York: Harper & Brothers, 1911), 121.

21. Robert Montgomery Bird, *Peter Pilgrim: or, A Rambler's Recollections* (Philadelphia: Lea & Blanchard, 1838), 529–30.

22. Ibid., 530.

23. Lydia Maria Child, *Letters from New York* (New York: C. S. Francis, 1845), 267.

24. Ralph Seymour Thompson, *The Sucker's Visit to the Mammoth Cave* (1879; reprint, New York: Johnson Reprint Corporation, 1970), 109.

25. Ibid., 126–27.

26. Horace C. Hovey, with Richard Ellsworth Call, *Mammoth*

Cave of Kentucky: An Illustrated Manual (1897; reprint, Louisville, Ky.: John P. Morton, 1912), 120.

27. Nathaniel Parker Willis, *Health Trip to the Tropics* (New York: Charles Scribner, 1853), 180–81.

28. Henry David Thoreau, *Walden* (Boston: Ticknor and Fields, 1854), 122.

29. Bayard Taylor, *At Home and Abroad* (New York: Putnam's, 1855), 214.

30. Ralph Waldo Emerson, "Illusions" (1857), reprinted in *Essays for Our Day,* ed. L. B. Shackleford and F. P. Gass (New York: Norton, 1931), 62–63.

31. John Muir, *A Thousand-Mile Walk to the Gulf* (1916; reprint, San Francisco: Sierra Club Books, 1991), 12.

20. Celebrities Underground

1. Joie Puskarich Shisler, Mammoth Cave National Park guide, email message to the author, May 11, 2011.

2. Joy Medley Lyons, Mammoth Cave National Park guide, personal communication with the author, July 2011.

3. Ibid.

4. Ibid.

5. Alan Sizemore, Mammoth Cave National Park guide, personal communication with the author, May 2011.

6. [George Prentice], "Mammoth Cave," *True American* (Lexington, Ky.), July 1, 1845. According to Nathanial Parker Willis in *Health Trip to the Tropics* ([New York: Charles Scribner, 1853], 175), George Prentice from the *Courier-Journal* was on the trip.

7. "Bull, Ole," Violinman.com, n.d., at http://www.violinman.com/Violin_Family/performer/Violinist/bull/0001.htm (accessed May 2011).

8. Willis, *Health Trip to the Tropics,* 175.

9. Ralph Waldo Emerson, "Illusions" (1857), reprinted in *Essays for Our Day,* ed. L. B. Shackleford and F. P. Gass (New York: Norton, 1931), 62–63.

10. Charles Shreve, "Jenny Lind," unpublished article, n.d., Mammoth Cave National Park Manuscript Collection, Mammoth Cave, Ky.

11. Janet Bass Smith, "Sable Melodists and the History of Minstrelsy," 2012, unpublished paper, Mammoth Cave National Park Manuscript Collection.

12. "Royalty Underground." *Louisville Courier-Journal,* February 2, 1872.

13. "Lt. Col. George Armstrong Custer Stationed in Elizabethtown," Hardin County History Museum, n.d., at http://www.hardinkyhistory.org/custerhere.pdf (accessed April 2011).

14. "When the Emperor Visited Mammoth Cave," *Ohio County News,* December 2, 1927.

15. Earle Dickey, "Cave City, Gateway to Kentucky's Cave Region," *Louisville & Nashville Employes'* [*sic*] *Magazine,* 1935, 13–14.

16. John Muir, *A Thousand-Mile Walk to the Gulf* (1916; reprint, San Francisco: Sierra Club Books, 1991), 12.

17. Quoted in Edwina Booth Grossmann, *Edwin Booth: Recollections by His Daughter* (New York: Benjamin Blom, 1894), 46–47.

18. "An Interview with John Nelson," originally recorded in the 1940s on cassette tape by Daran Neff, copied onto CD by Chuck DeCroix, Mammoth Cave National Park Audio Collection, Mammoth Cave, Ky.

19. Ibid.

21. *Colossal Cave Adventure*

1. Mary Bellis, "Computer and Video Game History," About.com, n.d., at http://inventors.about.com/od/timelines/fl/The-History-of-Computer-and-Video-Games.htm (accessed October 2012).

2. Stanley Sides, "Early Cave Exploration in Flint Ridge, Kentucky: Colossal Cave and the Colossal Cavern Company," *Journal of Spelean History* 4 (1971): 63.

3. Rick Adams, "*Colossal Cave Adventure* Page," n.d., http://rickadams.org.adventure/a_history.html (accessed October 2012).

4. Quoted in Dennis Jerz, "Somewhere Nearby Is Colossal Cave: Examining Will Crowther's Original 'Adventure' in Code and in Kentucky," *DHQ: Digital Humanities Quarterly* 1, no. 2 (2007), at http://www.digitalhumanities.org/dhq/vol/001/2/000009/000009.html.

5. Will Crowther, email message to the author, October 2012.

6. Jerz, "Somewhere Nearby Is Colossal Cave."

7. Crowther, email message to the author, October 2012.

8. Ibid.

9. Ibid.

10. Rick Olson, personal communication with the author, October 2012.

11. Quoted in Rick Adams, "*Colossal Cave Adventure* Page," n.d., at http://rickadams.org/adventure/c_xyzzy,html (accessed October 2012).

12. Jason Scott Sadofsky, dir., *Get Lamp,* documentary (Waltham, Mass.: Bovine Ignition Systems, 2010).

Index